Fernald Library
Colby-Sawyer College
New London, New Hampshire

Presented by

EDWARD L. ISKIYAN

In at the Beginnings

Philip M. Morse

In at the Beginnings:
A Physicist's Life

Philip M. Morse

The MIT Press
Cambridge, Massachusetts, and London, England

QC
16
M66
A34

Copyright © 1977 by
The Massachusetts Institute of Technology

All rights reserved. No part of this book may be reproduced in any form or by any means, electronic or mechanical, including photocopying, recording, or by any information storage and retrieval system, without permission in writing from the publisher.

This book was set in I.B.M. Composer Baskerville by Jay's Publishers Services, Inc., printed on Finch Title 93, and bound in G.S.B. S/535 by The Colonial Press, Inc. in the United States of America.

Library of Congress Cataloging in Publication Data

Morse, Philip McCord, 1903-
   In at the beginnings.

   1. Morse, Philip McCord, 1903-   2. Physicists—United States—Biography. I. Title.
QC16.M66A34   530'.092'4 [B]   76-40010
ISBN 0-262-13124-2

Dedication

This book is dedicated to the memory of Frank Porter, Jason Nassau, Dayton Miller, and Karl Compton, who had the most formative influence on the life I here relate—excepting, of course, that of my mother and father. Thanks go to my wife Annabelle and my son Conrad and daughter Annabella, who helped and at times suffered from the effects of my activities. Thanks also to the many friends who helped correct most, I hope, of my errors in recollection of things past.

Contents

Introduction
1

1
Germination
6

2
Education
33

3
Exploration
56

4
Fruition
92

5
Consolidation
118

6
Application
156

7
Invention
172

8
Initiation
213

9
Instigation
262

10
Promotion
309

11
Rumination
343

Name Index
371

1. Graduation from Lakewood High School, Lakewood, Ohio, 1921.

2. Newton Society of Lakewood High School, 1921. *Third row, left:* Frank Porter; *right:* Richard L. Barrett. *Second row, third from left:* Arthur Coffinberry.

3. Graduate school, Princeton University, 1929.

4. Senior members of Operations Research Group, U.S. Navy, 1945. *Second from left*: Maurice E. Bell; *third*: William J. Horvath; *fourth*: George E. Kimball; *fifth*: John R. Pellam; *sixth*: J. Pomerantz.

5. Dedication of Brookhaven Laboratory, 1947. *Speaker:* Eldon Shoup, vice president, Associated Universities, Inc. *Seated, from left, front row:* Lincoln R. Thiesmyer, Philip M. Morse. *Back row, left:* Lawrence A. Swart. *Behind speaker:* Mervin J. Kelly, representative, Atomic Energy Commission.

6. First meeting of Brookhaven Laboratory Board of Trustees, 1947. *Clockwise from front, second:* J. H. Van Vleck; *third:* Jerrold R. Zacharias; *fourth:* Frank Long; *seventh:* I. I. Rabi; *eighth:* Philip M. Morse; *ninth:* Eldon Shoup; *tenth:* Edward Reynolds, president, Associated Universities, Inc.; *thirteenth:* George B. Pegram; *sixteenth:* James R. Killian; *seventeenth:* Milton G. White.

7. Members of the Einstein Committee of the Atomic Scientists, 1948. *Back row, left to right:* Victor F. Weisskopf, Leo Szilard, Hans A. Bethe, Thorfin R. Hogness, Philip M. Morse. *Seated, left to right:* Harold C. Urey, Albert Einstein, and Selig Hecht.

8. Playing war games at RAND, 1955. *Clockwise from left*, John D. Williams, Philip Moseley, Lee DuBridge, Philip M. Morse, and Robert F. Bacher.

9. Governing Board of the American Institute of Physics, 1955. *Left to right, standing:* E. Rogers, A. V. Astin, R. F. Bacher, J. W. Buchta, H. D. Smyth, W. F. Meggers, H. A. Bethe, W. H. Markwood, Sr., S. A. Goudsmit, R. F. Patton, H. F. Olson, P. M. Morse, W. Shockley. *Seated:* M. W. Zemansky, R. A. Sawyer, F. Seitz, D. B. Judd, H. S. Knowles.

10. First Hellenic Operations Research Council, 1964. *Front row, from left, third*: General R. Spannyonakis; *fourth*: Philip M. Morse; *fifth*: Professor Papas.

11. Ceremonial dinner at Japanese Union of Scientists and Engineers, Kyoto, 1965. *Clockwise at table*, Philip M. Morse, Charles Salzmann, G. Murray, and Kazuo Tada.

12. Chairmen of the Board of the American Institute of Physics, 1975. *Left:* Horace R. Crane; *right:* Philip M. Morse.

In at the Beginnings

Introduction

A life, while being lived, has little pattern to it. If we are honest with ourselves, our actions seem to be erratic responses to the random torrent of events, less directed by any fixed intent of ours than by subconscious whims and by the peculiar sequence of circumstance. Only when a life is nearly over can one hope to find a pattern in the whole, not a steadfast purpose nor a transcendent reason but perhaps a vague outline of structure. A life must be viewed from a distance for the structure to suggest itself, as with the Monet painting of the Rouen cathedral at dusk. I have always liked to search for patterns in things; it may be I can find one in my own life.

Change surely is part of the pattern. Things have changed more rapidly, in my seventy-odd years of existence, than they ever have before—particularly in this country. Nearly everything we do differs from what people did a century ago. I can get to the start of a mountain trail in only thrice the time it takes me to get to my office. I can work most of the day in Minneapolis and yet be back home in Boston in time to witness on television the latest riots in the Middle East. Other countries are not quite as transformed as ours, but the change is apparent all over the world. Others may scorn our waste and our materialism, but they embrace the family car, the refrigerator, and the

In at the Beginnings

television set as fast as they can afford them. Whether we like it, whether we shall like it, the change is irreversible.

Scientific knowledge has caused much of the change—not the knowledge itself, actually, but its applications, too often uncoordinated and hasty. Some of the applications have been salutary, some precarious.

The big change, abetted but not caused by our new knowledge, is, of course, the geometric growth of populations, a growth now beginning to strain this earth's resources. In 1630, North America was infected with a few thousand energetic, prolific white men. By 1700, there were several hundred thousand, beginning to be too many for the East Coast. By 1800, five million people lived in the United States; by 1850, there were 23 million, too many for the country east of the Mississippi. The exponential explosion was on; 75 million by 1900 and 150 million by 1950. For the generation born about 1900, most natural dangers had been eradicated, the land had been altered but not yet dangerously poisoned. There still was room. But what about the generation born about 1950?

The knowledge that made so much of the change possible has also been expanding at a geometric rate. Until this century, the United States had little part in advancing scientific knowledge, now we contribute the largest share to its rising flood. I took part in America's scientific coming of age; perhaps my story will reflect a bit of its pattern of growth.

Pattern or not, of course, each life is unique. Mine certainly differs from most lives of my generation in the United States. When I grew up, very few in this country had turned away from practical affairs to follow an unremunerative career in pure science. In fact there was little realization that pure science was needed; engineers and inventors were the esteemed ones. Only with World War II did we discover that we could no longer depend on Europe for new knowledge to power our progress. I have seen, during my life, a little of how this discovery came about.

Even if there is a pattern in my story, however, why should it interest anyone but me? The life story of a scientist—an average scientist, not an Einstein—may not lessen the layman's uneasiness about scientists, may not persuade nonscientists to learn more about quantum theory or genetics or plate tectonics. But at least nonscientists reading this story can learn that such a life is fun to live. As for scientists, my tale of past struggles for freedom of curiosity or for more systematic forethought in applying new knowledge may tend to discourage some of my younger colleagues, who are fighting the same contests all over again. It should not, for these battles have to be fought anew by each generation, and our earlier tactics reported here may still be effective.

Most of my experience has been at the second, rather than at the top, level. I have seldom been in on the making of a big decision, nor have I made a major discovery. I have participated in many beginnings, however; have helped to exploit some new scientific breakthroughs; have aided in setting things up so that some big decisions could be made; and have been instrumental in carrying out some of the actions implied in their making. I have been closer to the grass roots than have those at the top level; I have therefore been able to see the inevitable contrast between what the decision-makers thought they ordered and what actually took place.

And I have watched the growth of the links coupling basic research and final application, links that must be more generally understood if we are to avoid other, grosser misapplications. Until recently, few scientists felt any more responsible for the way their discoveries were applied than workers on the assembly line feel responsible for automobile casualties. Lately, however, scientists (along with many others) have begun to worry about the way their discoveries are put to work. But they, like their critics, are frustrated by the task of changing the march of "progress." Where they differ from some of their critics is in their belief that more, rather than less, comprehension of science, by management, by government—in fact, by everyone—will help. They believe that if industrialists, legislators, government administrators, and other citizens understand more of the balance between the promises and the dangers of the bright new applications they (and not the scientists) put to work, perhaps this land will not become mortally ill. It won't do simply to ban "dangerous" research. If there is anything my life shows, it is that we never know what application may grow from an investigation and that all research has potential dangers as well as potential blessings.

Official histories of the beginnings of actions or discoveries are usually more orderly and logical than the events themselves. No narration of a beginning can possibly include all the hesitations and missteps, all the dramatis personae, all the contestants in each issue, or the complete sequence of actions from conception to consummation. Each participant has his own story of how the event started, seen from his own worm's-eye or bird's-eye view. My own recollection of the haphazard course of

Introduction

some beginnings of events of both major and minor eventual significance will often disagree with the official story and will probably contradict others' memories. I hope my tale of these beginnings will interest the reader not because it is more nearly "true" than the others but because it does present details of what happens when a spark begins to become a flame.

1
Germination

They told me I was born on August 6, 1903, at three in the morning; I don't remember. My seventy-year memory tape is a series of vividly recollected scenes, separated by blanks later filled in with conjecture and hearsay. The early scenes are disconnected flashes, glimpses of a now unfamiliar world, seen through a stranger's eyes. It takes effort to remember how different that world was, how many differences there are between the Midwest of 1910 and the East Coast of the 1970s. In Cleveland in 1914, when I was eleven, we had electric lights and streetcars and paved streets, but just outside the city the roads were of dirt and the farmers used lanterns and candles. No one had even a radio, much less a television set; our news of the world came days or weeks late and was that much less importunate. Pennies could actually buy things, not just pay the sales tax. Mail was delivered twice a day, and passenger trains were comfortable, speedy, and frequent. Cities were small enough so that people could get to work by streetcar or bicycle or even on foot in a half hour or less.

Horses pulled the icemen's dripping wagons, the junk carts, and the beer trucks. Horses were everywhere; automobiles were the capricious minority. In my grandfather's day, most people had their own horses, and my father was still experienced in their needs and foibles. My generation was the first that was not symbiont with the horse. My

Germination

family didn't own one, although we occasionally hired one when we went to our cottage on Lake Erie.

An early fragment on my memory tape is of roller-skating on the street in front of our house. The street was East 82nd Street, off Cedar Avenue—part of a slum district now. It was then newly paved with beautifully smooth asphalt ideal for skating. The horse-drawn vehicles delivering milk, groceries, ice, or coal were easily avoided. I don't remember ever meeting a moving automobile as I raced the length of the block. Our major perils were the sparrow-encircled horse droppings; if not seen in time they could lead to skinned knees or torn shirts.

Seen through a stranger's eyes; there is a vast difference between the me-now and the me-then. I find it as hard to understand my remembered, adolescent actions and emotions as I do those of my acquaintances of that age today. Some of those early memories are still vivid and visual. I'm not sure why they remain so near the surface, nor why, of all the myriad scenes I witnessed, they chance to be so carefully engraved.

Many of my "snapshot" memories have to do with the use and misuse of horses. For example, I can call to mind a picture of Cedar Avenue during a snowstorm, with horses clomping along, pulling wagons and carts and trucks. Beside me is another boy—I guess we were coming home from school. He throws a snowball across the avenue, and it happens, by pure chance, to hit one horse squarely in the eye. The horse rears, slips, and nearly falls. The driver, who has not seen the snowball, whips the horse and gets it going again. The boy beside me gives a whoop of laughter—and then gulps. And that is the end of the scene. I can play it over any time I wish—but does it tell me anything?

Many of my less vivid memories have been generated from reading or from school or from parental discourse; these are second-hand memories. We had a lot more evening conversation then. My father read a lot and liked to read aloud—Dickens, Mark Twain, stories from the *Saturday Evening Post*. And these would often start him telling of his own memories and of stories he had heard from *his* father. I got a piecemeal picture of three generations of technical men, builders, and planners, with each generation taking part in the new technological development of the time.

My great-grandfather, John Flavel Morse, had come to Ohio from Massachusetts as a boy in 1813. His father helped to found the town of Kirtland, east of Cleveland. When John grew up he became a builder of houses and also a social activist. He opposed Joseph Smith's Mormons, who tried to take over Kirtland in the 1830s. Later he was an abolitionist, helping run the underground railway bringing runaway slaves out of the South to freedom in Canada. He was elected to the Ohio legislature as one of the first "Black Republicans." In the 1860s he worked for the federal government, designing and building post offices and customs houses all over the country. The house that he, his father, and his brother built near Kirtland was still in the family when I was young. We would visit it during our family reunions on the Fourth of July. I can remember admiring its oak beams and the wooden pegs that held them together.

My grandfather, Benjamin Franklin Morse, surveyed railroad rights-of-way during the era of rapid rail expansion after the Civil War. He laid out what became the main line of the Nickel Plate Railroad across Ohio, and he assisted the famous engineer Charles Collins to put together the New York Central extension into Michigan. But the arbitrary ways of the railroad magnates disgusted my grandfather so much that he quit the railroad to take the job of chief engineer for the city of Cleveland. Those were the times of Cleveland's big expansion, including Mayor Tom Johnson's "socialist" plans for a city-owned street railway, the creation of an extensive system of city parks, and the construction of high-level viaducts across the Cuyahoga River valley.

I remember my grandfather after his retirement; he was short, heavy-set, taciturn, with a mop of wavy gray hair and a mammoth mustache. I understand he was a family tyrant. I don't remember my grandmother, who died when I was eight. She once confided to my mother that the grief she had experienced at her own mother's death, when she was young, had been so great that she had sworn to spare her children this sorrow, by taking care that they did not love her too deeply. It must have been a rough family life for my father and his two brothers and two sisters.

My father was the youngest son and evidently somewhat of a problem to his family. After high school, in 1893 he went to the newly founded Case School of Applied Science for two years, dropped out to work surveying the Cleveland park system, and then worked in telephone-system construction, that era's glamour technology. He supervised the building of several telephone exchanges in Ohio, including one in East Liverpool on the Ohio River, where he met my mother and married her in 1901. He worked for many independent telephone companies and, when most of these gradually were absorbed by the

Bell system, he became a consultant for various state utilities commissions, assisting in their fight to control telephone rates. He would be away for months at a time and then would be at home with nothing to do but read, or work slowly and meticulously in his well-equipped shop, making a chessboard or a table. He hated big business, as exemplified by Bell, a hatred perhaps initiated by his father's hatred of the rail magnates, but certainly strengthened by the antitrust movement then evolving. This hatred eventually cramped his career and affected his family.

My father was a quiet man, bullied by his father and, later, sometimes by my mother. His anger could be aroused, though. An argument he once had with my uncle about Theodore Roosevelt nearly led to blows halfway through dinner. A more vivid memory flash comes from an episode at our summer home on the shores of Lake Erie. We arrived one summer to find that a tug and barge were regularly anchoring offshore to remove sand, presumably for sale elsewhere. They had already sucked away half our beach.

My memory of that episode is as immediate and reproducible as a movie reel. There were the tug and barge, motionless in the calm water. The sucker pipe, like a huge heron's neck and head, had its bill sunk to the bottom. I can still faintly hear the slow beat of the pump as it sucked up our beach sand. And there was my father, rowing back to shore after a futile argument with the tug captain. He beached the rowboat and, without a word, marched up into our cottage, soon to return, red-faced and still wordless, with his army-surplus rifle. He fired only two shots.

Germination

At least one went through the sucker pipe; he was a pretty good shot. Then, finally, there was sudden action aboard the vessels. The heron bill was raised, and the tug and barge headed out. They never came back.

The land so resolutely defended had been acquired by my father and his brother around 1900. It lay along the shore of Lake Erie, just east of the Chagrin River and less than twenty miles from downtown Cleveland. The twenty acres of land had nearly a mile of sand beach and a few acres of fields. The rest was woodland, containing a few huge black walnut trees, relics of the original forest cover. The nearest town on the railroad and interurban trolley line was Willoughby, four miles away by dirt road. Each brother built his family a cottage not far from the shore. The fathers would come out to the cottages on weekends and during their vacations, but the mothers and the children stayed there the whole summer. We never had electricity: food was kept from spoiling in an underground cool box; water was pumped, first by hand and later by a one-cylinder engine, to a pressure tank from a pipe sunk in the sandy soil.

We were almost completely isolated at the lake. The road from Willoughby was back of the woodlot and there wasn't much traffic on it anyway, so the most pervasive sound was the murmur or crash of the waves. We raised many of our vegetables. Twice a week a horse-drawn wagon stopped by with meat and canned goods. Once a day my cousin John and I would make the two-mile round trip to the nearest dairy farm to bring home the daily milk supply in four half-gallon pails. When the weather was dry we would bicycle, raising clouds of dust and occasionally getting a wheel caught in ruts, spilling ourselves and the milk. In rain we would slog it on foot—barefoot, of course, to save our shoes. I can still taste the milk, cream-rich and warm; the present homogenized and pasteurized liquid is a quite different product. Nor do today's mammoth strawberries have the flavor of the tiny wild strawberries we would pick alongside the road, their perfume enhanced by a trace of road dust. Even now, when I remember their taste, I can feel a bit of grit between my teeth.

At least half our time was spent on the beach, much of it in the water. Lake Erie water was pure then, nearly pure enough to drink even without being filtered through the sandy soil. And the sand was not littered with bottles and beer cans, but only with a few dead fish and lots of driftwood, enough to keep us in firewood for the chilly days.

My mother used to say that my outdoor life each summer kept me alive during the winters. Each winter I managed to be out of school for nearly two months. The usual children's diseases didn't bother me. My sister Louise was once extremely ill with measles; all I had was a mild rash. I went in for fancier ailments: streptococcus infections, mastoiditis, appendicitis, and the like. Before antibiotics these could be deadly. My troubles didn't come from a lack of proper food. Mother was one of the early enthusiasts of the balanced diet. We ground our own wheat and had plain cod-liver oil each day. I can remember my brother Allen once getting a spoonful of peppermint-flavored cod-liver oil by mistake—he spat it out and grumbled something about spoiling the taste of good cod-liver oil.

My mother took all our illnesses in her stride; she was both energetic and ambitious. Her father, William McCord,

Germination

had been the editor of the East Liverpool, Ohio, newspaper. Mother, when she finished high school, became the paper's girl reporter and worked at it enthusiastically until my father married her—or she married him. She made sure all her five children got an education. All of us had music lessons, and all of us were put through college, no matter what the state of the family budget. We also all went to the nearby Presbyterian church, sang hymns, and learned the books of the Old Testament.

My lifelong friend, Dick Barrett, writes, "Your mother, in particular, was a very remarkable woman and your father was a very lovable man." It could be. I cannot judge; I was too close to them.

I worked at my violin lessons for five years. I did learn to read music and, later, in college, I enjoyed playing occasionally in an amateur quartet. But my practice never made me perfect. Mother was more successful with my brother Richard. At the age of five he decided he would learn to play an old bugle he found in the summer cottage, and he tormented us with his efforts all that summer— except for the three days it took him to remove a cork my cousin John and I had jammed down the bugle's throat. Dick went on to become a trumpet player, earned enough by his playing to put himself through the Eastman School of Music, and eventually became head of the music department at Ripon College in Wisconsin.

I got along well in school; in spite of being out nearly two months each winter, my grades were mostly A's. The lessons couldn't have excited me much, however; my most vivid memories of that time are all extracurricular.

Mine was the first generation to learn in its youth the

care and control of the automobile instead of the horse. I was fifteen when I first had a chance to drive. Given the lack of competing traffic, the model T Ford was ideal to learn on. Once one got through the intricate ceremony of priming, cranking, and warming up the engine, the patient beast would respond to every change in the controls—all one had to learn was which change resulted in the desired response. On icy days I would go to a wide, nearly deserted street and practice skidding, learning when to turn the front wheels, when to apply power, and when to "leave her in neutral." The final test of skill was to make a U-turn skid across the road, coming smoothly to rest, parked at the curb and pointing in a direction opposite to the initial motion. The ability to control skids, once learned, is a permanent skill, like riding a bicycle. I think I can still control skids better than most drivers supposedly in control of their superpowered, automatic-gear-shift juggernauts.

While I was still in grade school, I read voraciously on my own, whether I was sick or well. I read through my father's large collection of Dumas, Dickens, Stevenson, Kipling, ancient history, and the *National Geographic*. And I devoured the collection of the local public library: H. Rider Haggard, Henty, Mark Twain, more history and exploration. I avoided Scott and Longfellow, because they were analyzed to death in school. I was out of college before I lost my aversion to Shakespeare, induced by dreary recitations of *Julius Caesar* and *The Merchant of Venice*. Much poetry still leaves me unmoved.

Chemistry attracted me. My closest friend, Milton, was the son of a local druggist. We collected a variety of chemicals: sulfuric, nitric, and hydrochloric acids, ammonia, and potassium hydroxide, none of which would be sold to a schoolboy now. My father set up a laboratory in the basement, with a sink and a ventilating hood. I don't remember any serious accidents with our lethal assortment. We did blow up a batch of gunpowder we were drying, incautiously, over a Bunsen burner, but the explosion took place in the hood, so all we got was a scare. We also memorized the table of elements and learned the nature of chemical reactions.

Isolated facts didn't interest me much; patterns in facts were what excited me. I didn't bother to memorize individual chemical reactions; the Mendeleev table of the elements, with its recurrent pattern of chemical properties, was what I worked to understand. The unifying concepts of evolutionary theory entranced me, and I ate up the few books on geology and paleontology in the library. I remember with some shame my attempts to persuade my grandmother McCord that man must have come from the apes. She, of course, believed literally in the Bible, and it must have been a strain for her to hear a favorite grandchild spout such impiety.

Those broad patterns in science that caught my young fancy were mostly then just heuristic assumptions. Their underlying reasons have only gradually been discovered since then: atomic structure to explain the periodic table, Mendelian genetics and then molecular biology to elucidate evolution. But the patterns have remained. I once went through a period of deep dejection during my grade-school years. I began to be afraid that everything in science had already been discovered, since the texts and other books I read presented science as a completed whole. I enjoyed the stimulus of reading about others' discoveries—

I still enjoy that kind of second-hand exploration. But I wanted to find something new, all by myself, and everything seemed to have been discovered. Since then, of course, more has been discovered than was known at that time.

I must have been a solemn and retiring youngster, managing to avoid most battles and arguments. I was clumsy and not much of an athletic competitor. In one mass schoolboy battle I remember thinking, "This is silly," and walking away from the melee, untouched and presumably unnoticed. Once, in a fit of spring madness, I tackled Eddy, the class athlete, in a wrestling contest. It soon ended in a bloody nose for me and profuse apologies from Eddy; we remained friends.

The seventh- and eighth-grade boys had to learn wood- and metal-working. I was never very good at it, but I relished it. I still use my wood-turning tools today with that exciting mixture of pleasure at seeing the projected form take shape beneath the flying chips and of fear lest the tool slip, gouge, and ruin the piece—it often did and it sometimes still does. I enjoy that same mixture of achievement and fear while rock-climbing.

In 1916 I went on to Central High School, in the older part of Cleveland's east side. All the races and nationalities that had piled into Cleveland were represented: Italians, Germans, Hungarians, Jews, blacks, Chinese. My Latin teacher was black, and an admired classmate was a Polish Jew. The only nationality unrepresented were the Irish, self-segregated in their parochial schools. I then knew only one Catholic, a boy who lived just around the corner. He

Germination

and I got along well, but his family's clannishness strengthened the antipapal prejudice I had absorbed from my father's New England traditions and from my mother's Scotch-Irish Presbyterianism.

But by high school I had begun to form my own prejudices. It took only a month of gym classes for my dislike of regimented calisthenics to bring me to the point of exploring alternatives. The physical education teacher was of the usual energetic, no-nonsense type, but he evidently had a sense of humor—or of sympathy. I don't remember what sort of argument I used on him, but I was admitted to a special group training for a gymnastics demonstration to be put on in the spring. From this distance, my satisfaction with the transfer seems hard to understand, for the special group demanded much more time and work than the regular gym classes. And my clumsiness drew the wonder and derision of the rest of the group. I managed to learn the forward roll, but I never did master the back flip and my performance on the low bar was anything but graceful. Nearly always, when I was third man high in a pyramid, the whole thing would collapse. Why I wasn't booted off the team I don't know; perhaps the coach felt I was a challenge. Besides, I could play the violin, and this was to be part of the program. Clumsy as I was, I did learn to handle my body, which helped later when I began mountain climbing. I still can control myself—landing safely after a fall, for example—better than most.

The second year of high school was 1917, and everything was war and soldiering. This time I couldn't evade drill; every student had to wheel in squads and learn the manual of arms. Sample trenches, complete with barbed wire, scarred the central square in Cleveland. My Boy Scout troop, when it wasn't parading, was selling Liberty bonds. At one meeting my Scoutmaster told me, somewhat conspiratorially, to call at a particular address that evening and go through my sales pitch. It turned out that I didn't have to say much; when I arrived the resident had a pledge for $10,000 worth of bonds already made out. As a result of that prearranged sale, I sold more bonds in that drive than anyone else in the troop. I was awarded an impressive medal, which I wore on my Scout uniform with some pride, mingled with embarrassment at a too-easy win.

Because of the war, my father had no more consulting jobs; he had to return to work for the local independent telephone company, in charge of maintenance, a position he had left some years before. That autumn the influenza epidemic closed all schools. My father got me a job in the central telephone exchange, servicing the big bank of emergency batteries. I spent most of the time listening to my boss, a veteran power lineman, reminiscing about the old, dangerous days of electric-power transmission and predicting that all the independent telephone companies, like the independent power companies, would soon go under.

At about this time we sold our house and moved across town to the suburb of Lakewood. I finally caught the flu about the time we moved, so I don't remember much about it. Financial stringency also meant that the cottage by the lake was sold, so the summer of 1919 I was sent to a curious nautical-military camp, run by the Lakewood Yacht Club on its island at the mouth of the Rocky River. There we campers drilled and paraded and slept in army tents, but we also swam and sailed. I made some friends

Germination

among the other boys and also with a young ex-Navy radio operator, Dick Roberts, who taught us a bit about radio circuits and drilled us in Morse code. Radio transmitters then used rotary spark gaps, and radio receivers generally consisted of headphones and galena crystals. Vacuum tubes were mysterious rarities, and radiotelephony was a laboratory toy. If you wished to join the ranks of the radio amateurs you learned code, and you kept your family awake nights with the noise of the spark gap.

During my last two years of high school, at Lakewood High, I began to grow up intellectually; I find I can begin to understand some of my remembered thoughts. I still kept pretty much to myself, and I was still sick for several months each winter. My reading expanded: I read H. G. Wells, particularly his *Outline of History,* which extended my understanding of history and prehistory. I read Darwin and Jacques Loeb and Osborn's *Men of the Old Stone Age.* I spent hours of concentrated effort drawing up a thirty-foot-long temporal chart, synchronizing the histories of all the nations with their rulers, from Menes to Napoleon, from Sargon to K'ang-hsi. I spent days learning how to letter and illuminate pages in the style of the Book of Kells. I read Breasted's *History of Egypt,* which made me want to be an archeologist. But then I read E. E. Slosson's *Creative Chemistry*, and I again decided to become a chemist. Almost all the science I read about then has since been revised, several times in some cases. I like it that way; it means that when I return to ancient history or evolution or cosmology, each has become a new subject, with new concepts and mysteries to excite me all over again. If I can't break new ground myself in all these fields, my explorations at second hand continue to be stimulating.

In at the Beginnings

I never wanted to be a mathematician, perhaps because mathematics was always taught as a tool, not as a field for intellectual exploration. Geometry, it was implied in high school, had been completed by Euclid, and more interesting subjects, such as number theory or the theory of equations, were too recondite for tender young minds. Perhaps it was just as well that Eric Bell's *Men of Mathematics* had not yet been written; if it had, I might have got started on the mysteries of prime numbers or Diophantine analysis, and my whole life would have been different. Conceivably, my love for pattern and symmetry would forcibly have led me deep into congruences and quadratic residues, but by the time I finally learned about number theory I was too busy with other things.

I began to notice girls, although not yet with any degree of concentration. They were still mysteries, even my sisters and my cousins, and they all seemed to have a view of the world quite unrelated to mine. This was the era of short skirts; somehow the girls' silk-clad legs asked to be looked at, although you knew you were not supposed to stare. And this seemed illogical too.

In 1920 I was asked to come back to summer camp as a junior counselor. I went out four days early to help set up the camp. My tentmate, another junior counselor, was an imperviously cheerful lad, eternally chewing gum and interested in any kind of sexual activity. Our three nights by ourselves before the camp opened were educational for me. The second night I was rather pressingly invited to cooperate in an exercise that I did not understand at first and then thought rather unhygienic when I did understand. Refusal did not dent my tentmate's cheerful friendliness. The next evening he took me along on a double date

Germination

with a pair of girls he had already managed to locate. I'm afraid my participation in the evening's activities went little beyond watching the other pair's performance, but, as I said, it was educational.

My friend Dick Roberts, the radio operator, was back that second summer, too. He was full of plans to start a store in downtown Cleveland, selling supplies to radio amateurs. Commercial radio broadcasting had not yet begun, but many radio operators trained during the war wanted to build and operate their own transmitters and receivers. Roberts was convinced that supplying what they needed—earphones, variable condensers, wire for coils, power transformers, and the rest—would be a good business. Before the summer was out he had asked me to join him and two friends in the venture. I was seventeen at the time, and my knowledge of business was even more negligible than my experience in radio operation. I suppose Roberts and his friends needed more money to get started and hoped my father would contribute. Later that fall my father did just that, over mother's protests.

But in the meantime I entered my senior year at Lakewood High and began the one precollege course in science that truly caught my interest. I had taken beginning chemistry and physics earlier and had made good grades, but I had been bored by the pedestrian teaching, the slow rate of progress, and the crude laboratory experiments. The courses were taught as a set of rules and recipes, not as an intellectual adventure. I had glimpsed some of the adventure from my own reading, but had nearly given up getting any of it in school. My reaction was not to skip the classwork; it was rather to finish it as soon as possible so I could do my own exploring.

However, Frank Porter, the head of the Lakewood science department, offered an elective course in chemical analysis that I decided to take. It was a laboratory course with only four students. Porter never lectured in the course; he just dropped by and talked to us while we were busy at the bench. But the work and his comments about it gave me a feeling for the uncertainties, the disappointments, and the recompensing triumphs of scientific exploration. Our tasks were very simple. We were given a series of unknown solids or solutions and were to apply a standard set of tests to determine what they were. None of the unknowns were hard to identify, but Porter contrived to bring home to us some of the pitfalls in all research. His gentle "It seems to me, Philip, that you weren't very careful in that test" brought home the dangers of sloppiness better than an hour's lecture. And his slightly acid queries ("Are you sure? Have you tried a chloride?") kept reminding us not to jump to conclusions. In the years since, others who have made a name in science have told me they also were inspired by one unusual high school teacher, like Frank Porter.

The class also brought me a lifelong friend, Richard Barrett. I had met Dick the year before, but when we took the chemical-analysis course together, we became inseparable. He had grown up in a small town near Cleveland and had had to work hard for everything. He had to work nights to earn enough to go to high school. Later he would take three years off to earn enough to go through college. He eventually was offered a graduate assistantship and earned a Ph.D. in mineralogy from Ohio State and then a place on the faculty at Case Institute. But tuberculosis caught up with him and he had to move to the Southwest, where he ultimately became Dean of Arts and Sciences at New Mexico State University. I always looked up to Dick Barrett; he had had to grow up faster than I had; he always seemed to have the answers.

Another close friend was Arthur Coffinberry. An accomplished violinist, he was also interested in science and never was quite able to decide which career to choose. The three of us organized a student science society at the high school. I was elected its first president, although the other two did most of the organizing. My contribution was to deliver two overlong talks on special relativity, based entirely on Einstein's popular book on the subject, which I had just finished reading. I doubt that either the faculty members or the students who attended the talks got much out of them. I had to follow Einstein's text slavishly, and my mathematics was not up to the subtleties required. My deficiencies didn't hamper my self-confidence, but it was a number of years before I could begin really to understand the beauties of the theory.

That spring I had my last serious illness, which was perhaps a form of polio, although no permanent paralysis resulted. (And only twice in my life since then have I ever had any ailment more serious than a cold.) I recovered in time to finish my classwork and to deliver one of the four orations at commencement (one other was given by Dick Barrett). I learned that I had been admitted to Case School of Applied Science (later Case Institute of Technology, and now part of Case Western Reserve University). Both my father and my uncle Frank had gone to Case, and it had been assumed for some time that I would go there. I was satisfied. I wanted to follow my own path, but I had already begun to learn how I could do what I wanted to within the system. And, in addition, I was interested in chemistry. Case was across town from our home in Lakewood, so arrangements were made for me to board with an old acquaintance of mother's, whose house was near Case.

Meanwhile, Roberts and his two friends had, as they had hoped, established a small store, called the Radiolectric Shop, on the third floor of a downtown building, and I joined them for the summer. We managed to collect a stock of transmitting and receiving equipment, including even a few of the vacuum tubes that were beginning to come on the market, some of them war surplus, some of them made by Western Electric for use in the Bell system's long-distance lines. We began to attract customers and to meet their demands for esoteric equipment.

When commercial radio broadcasting began the following year, the craze swept far beyond the small clan of amateur operators we had known. Fortunately, our store had made contact with the right manufacturers and wholesalers and was ready to meet the craze when it came. But in that first summer we couldn't have anticipated all that. We just kept the store open and hoped we had enough in the bank to pay for the next shipment—and maybe to pay for lunch and carfare. And then in September I went on to Case, coming in to the shop only on Saturdays.

Case was then, for the most part, a training school for engineers and applied scientists, in the American tradition of practicality. It had a rigid curriculum, based on the correct assumption that its students came with a bare minimum of high-school science and with no knowledge of mathematics beyond geometry and a smattering of algebra.

Germination

In their freshman year all students had to take chemistry, mechanical drawing and surveying, trigonometry, and analytic geometry. As a gesture to the humanities we had to take two years of English and of a foreign language and one term of history. I didn't mind this sparse fare; I felt I could get the humanities by myself.

But the work of this first year began to bother me. The humanities courses were dull drill—inspiring teachers in these subjects do not immure themselves in a technical school, with no graduate courses to keep them on their toes. I found I already knew enough chemistry to pass the classwork without study. And the chemistry laboratory was pure misery, after the individual instruction I had received from Frank Porter. The unknowns we were supposed to analyze were sloppily prepared, and the reagents we were to use in analysis were impure. Every test gave some reaction; we were supposed to report only those that gave overwhelming results. The instructor was caustic and impatient. If this was chemistry, I was not sure I wanted to be a chemist.

The one bright spot was my mathematics instructor, Jason Nassau. He was something of an anomaly, both in his position at Case and in his attitude toward teaching. Case had an astronomical observatory, donated by Warner & Swazey, the machine-tool and telescope builders, but it had no astronomy department. When Nassau was engaged as director of the observatory, he was appointed assistant professor of mathematics—which was not entirely irregular, as he had a degree in mathematics from Edinburgh as well as one in astronomy from Syracuse. He taught the only class offered in astronomy, and the rest of his schedule was taken up by mathematics courses. He was interested in what he was teaching and interested in his students. I was lucky enough to be assigned to his section. I took to him immediately.

He had a high-pitched voice and an accent that I could not place, although I could diagnose most of the ones heard around Cleveland. Later I learned he had been born in Smyrna, presumably of Greek parentage. I do not remember how I aroused his interest but, by the second semester, I was a regular visitor to the observatory and to Nassau's office in the old Main Building.

There was plenty to do that winter of 1921. I worked on Saturdays and at other spare times at the Radiolectric Shop and began to learn my way around Cleveland. One of my fellow students was from Cleveland's Chinatown; another was from the city's large Hungarian section. From them I began to learn that eating can be an adventure as well as a pleasure. Dick Barrett and I joined the Cleveland Photographic Society, learned about composition and cropping and mounting, and struggled with the bromoil process.

In the Taylor Arcade downtown was a tiny book store run by Richard Laukhuff, which Dick and I began to visit regularly. Laukhuff had come from Germany before the war, but his accent and his manners revealed his Rhineland origins. His shop was completely un-American, in the best sense of that word. He would talk for hours about books or about special printings, and he would reach up to the top shelf to bring down, blow the dust off, and ceremoniously open one of his treasures to illustrate a point. It was not that he wanted us to buy one of these specimens; in

Germination

fact, he would argue against it. The popular books on the lower shelves were what he made his frugal living on, but he did not think much of the customers who bought them. He erased, for me, the picture of the brutal Prussian that the war propaganda had painted of all Germans.

Those were yeasty times in Cleveland. The Cleveland Playhouse was just getting started in a remodeled church, putting on Shaw and O'Neill and Pirandello for devoted audiences. Compared to its later productions, when it had earned its own theater building and had graduated some of its cast to Broadway, the work of the Playhouse then was less polished, but the excitement was high. One of my Photographic Society friends was helping with the scenery; through him I got acquainted with the whole enterprise, both in the audience and backstage.

My musical friend, Art Coffinberry, was a subscriber to the Cleveland Symphony Orchestra concerts, to which I went when he couldn't use his ticket. Suddenly I learned that music was more than an agonizing struggle to move the bow correctly and to place the fingertips accurately on the proper string. At that time the Cleveland Symphony played in Masonic Hall; it got its own Severance Hall later. The orchestra usually played the old reliables: Beethoven, Brahms, Franck, and occasionally Debussy. But this repertoire was what I needed to catch up on my education. I would follow the concerts with a borrowed score and would afterward try to understand Berlioz' book on orchestration. It was a new taste to explore, one that I still enjoy.

Those also were troubled times in Cleveland, as postwar hatreds and suspicions festered. Cleveland was a labor-union city, and lines were being drawn between the unions, recovering from their wartime acquiescence, and the conservative Republican upper class. The words Soviet and Bolshevik were beginning to be imprinted as epithets of fear and hostility. I can remember a May Day, near the center of town, first seeing squads of the newly formed Veterans of Foreign Wars marching toward the central square, all carying baseball bats and then, later, seeing a few bloodied victims being helped away by friends. This was the time of Attorney General A. Mitchell Palmer, the World War I counterpart of Senator Joseph McCarthy, who was making headlines with his spy investigations. Even my friend Laukhuff was questioned because he was German and sold foreign books and subversive magazines, such as the *New Republic*. My family prejudices against big business were beginning to be reinforced by my own observations. My father was a Bull Mooser and was vocal in his disgust at Harding's election.

By the spring of 1922 my family's finances were near crisis, and it was decided that I would have to stay out of Case for a year to work at the Radiolectric Shop and earn some money. I didn't mind. My first year at Case had not been exciting, except for Professor Nassau. A year away from school might rekindle my interest, or it might deflect it in a new direction.

In retrospect I feel sorry for the college students of the fifties and sixties. We in the twenties had much more freedom to take a steadying pause in our schooling than they do today; from 1940 to 1970 a boy stayed in school or he got drafted. Also the work force was much more permeable then. If he wished to spend a summer or a year working as a carpenter or a steward on a passenger ship to South America, he wasn't barred by some union's seniority rules. Practically any job was open to him if he was willing to work; I guess my generation was the last for which this was true. The pay wasn't much, of course, but that was less important than the feeling of freedom of choice we had. In the fifties and sixties, when I was student adviser at MIT, I saw many youngsters whose motivation was eroded by their inability to catch their breath for a year, by their having no choice but more school or the army.

I certainly picked the right year to take off from school. It was in 1922–23 that the radio craze hit Cleveland. KDKA, the Westinghouse station in Pittsburgh, had started broadcasting news and music the previous year. Other stations were starting up all over the country. Suddenly everyone wanted a receiving set to listen in with. Newspapers started carrying news about programs, as well as technical articles about headphones, crystal sets, vacuum tubes, and loudspeakers. Our store was one of the few with any knowledge of the new developments and any contacts with the sprouting radio industry. Our trouble was a lack of ready cash. We couldn't advertise enough to capture very much of the new market, and, even if we had, we couldn't have bought enough stock to satisfy the demand.

We did as well as we could. Many amateurs turned to making sets for their friends. When they came to us for parts, we persuaded some of them to build sets for us to sell. For a while we had a near-corner on vacuum tubes in the city, but we were too short on other equipment to build a line of audio amplifiers. Our stock changed almost every week; we would be out of one thing, while some other new product would just have come in. It was fun, but it wasn't the way to make a lot of money. The store simply couldn't support four partners. So first one and then another went off to other jobs, leaving Roberts and me to run things with some part-time assistants who, we hoped, knew more about vacuum-tube circuits than we did.

During the year the radio market lost some of its anarchy. Dependable receiving sets and vacuum-tube amplifiers became available. We could get loudspeakers that reproduced sounds more faithfully than the usual headphone attached to a horn. We began to build up a knowledgeable and helpful clientele. But we still had a problem keeping in stock all the new products they wanted. Other stores, with no technical knowledge but more capital, began to eat into our near-monopoly. Nonetheless we were making some money, and it looked as though I could afford to return to Case the following year. My short immersion in the marketplace had only strengthened my desire to learn more science, particularly the science related to radio. My father, in the lengthening intervals between his consulting jobs, had begun to help out in the shop. He would take my place when I returned to college.

During that parenthetic year I began to sort out my impressions of the sciences. Through Professor Nassau I had become acquainted with a studious upperclassman named Louis McBane. He was interested in radio, and I would meet him from time to time at the shop. McBane had chosen to specialize in physics and was enthusiastic about the program. At that time the field of physics was practically unknown; many people confused it with a branch of medicine. McBane claimed it was the fundamental science; its branches—dynamics, electricity, optics, heat, and acoustics—covered the basic elements of all natural phenomena. Newton and Galileo had been physicists, he said. Einstein was one. If one liked science, one ought to learn physics. Since I did want to explore all science, I began to think that maybe physics was the way.

I was sure I also wanted to experience the capabilities and pleasures of my own body. Investigation of exotic foods was one beginning. Greek, Chinese, Italian, Hungarian, and other restaurants were there to be sampled and compared. Cigarette smoking was becoming widespread, but there were still many pipe-smokers. I found a shop that had the basic tobacco varieties—burley, Virginia, perique, latakia, Turkish—and learned to mix my own pipe tobacco. I still do, although it is harder now to buy the elementary varieties. It was a small gesture against common fashion, a predictable response to the reading of Huysmans and Machen.

Alcohol was also a realm to explore, despite Prohibition. Northern Ohio had been a wine-producing region, and my family had always made wine and had a bottle of whiskey reserved for emergencies. But they were the exception; most of the parents of my friends viewed drinking as the prelude to moral downfall. This, of course, made it a matter of honor for many of the younger generation to explore alcohol thoroughly—liquor was our generation's marijuana. The alcohol most often available in Cleveland was cheap red wine, bought in the Italian section. Occasionally we could come by a quart or so of straight alcohol pure enough to drink, or so we hoped. We would flavor it to taste and dilute it to a proper proof, before using it for long drinks or amateurish cocktails. The easiest flavor to prepare was gin, and each of us had our favorite proportions of oils of juniper, coriander, and lemon, to add cautiously by medicine dropper to the spirits.

The conveniently located apartment of a hospitable friend of Dick Barrett, Gordon Williams, was the site of a number of parties. Gordon had a collection of orchestral records, 78 rpm's played with a mechanical pickup. Their tone quality would curl your ears nowadays, but they served then to start off an evening of arguments about composers and compositions, especially if Art Coffinberry were present.

2
Education

In the fall of 1923 I was back in college, as a sophomore. This time I was more certain as to what I wanted to be—or, rather, what I wanted to learn. My interest in science had, if anything, been strengthened by my taste of the world of business. I still wanted to explore, to learn, more than I wanted to manipulate people. I felt I could delve into history and literature, unaided, more easily than I could penetrate mathematics and the sciences by myself. My habit of reading—five books a week, as a rule—made possible an individual survey of the humanities that I did not feel able to achieve in the sciences, where drill and a teacher's guidance were necessary even to get started. My unguided tour through the humanities would likely be unbalanced, but I hoped that this could be made up for by the sheer number of books read. My proposed program also militated against my reaching any degree of fluency in other languages, but I already had little liking for classroom linguistics, as then taught.

My self-analysis also reinforced family tradition, which pointed toward continuing a science program at Case. The analysis probably was only a rationalization of a journey already begun, but I have never regretted it. My reading is still mostly in nonscientific subjects, and my interest in the humanities is still great.

When I told my father that I was electing the physics

program, his only comment was, "That's fine, but what will you do for money?" My choice of that obscure subject had to be made on faith, for the only physics course I had had in high school had covered a hodgepodge of simple mechanics and out-of-date electricity. Frank Porter, the only inspiring science teacher I had had in high school, had awakened my interest in chemistry, but my experience with chemistry in the first year at Case had convinced me that it wasn't "basic" enough for me. Louis McBane, who had gone on in physics while I had been at the Radiolectric Shop, continued to be enthusiastic about his choice and helped to persuade me to shift. I was due to take the first course in physics in the coming year; if I didn't like it, I could still change programs again in my junior year. I didn't share my father's worry about money. I was already attracted to a college teaching career. Most colleges had physics departments. Professors never got rich—but then they never seemed to starve.

By the end of that sophomore year all doubts had disappeared. Both major courses, physics and calculus, had stimulated me. The physics lectures were given by Professor Dayton C. Miller, the head of the department and one of the handful of eminent American physicists at that time. He was tiny and neat and polished, with an imposing mustache, beautiful white, wavy hair, and a pleasant but very formal manner. His lectures were clear but not theatrical; his lecture demonstrations were carefully planned and always worked. His exposition followed the English school of the previous century, avoiding calculus. In spite of the awkward mathematics I began to see how it all fitted together.

Miller's lectures on acoustics were particularly interesting. He was an expert flute player, and his collection of

Education

flutes, some of which he had made himself, was renowned. He designed and built the phonodeik, a mechanical recording device to trace the shape of sound waves. To analyze the magnified curves that the phonodeik produced he used a Henrici harmonic analyzer, which mechanically measured the strengths of the various harmonics in the graph of the sound wave. The phonodeik, being mechanical rather than electronic, introduced into the recordings many distortions that were impossible to correct completely, but Miller's work was among the first quantitatively to analyze musical sounds, instrumental and vocal. Electronics, with its marvellous ability to magnify weak signals faithfully more than a million-fold, lets us forget how difficult it was to measure sound mechanically.

Miller demonstrated some of his work in his sophomore lectures and indicated how the work could illuminate the relationship between the structure of a musical instrument and the sounds it produces. Into these lectures there crept a hint, which I had missed in all previous science courses, that not all was known, that still more was to be learned.

The course in calculus also stimulated me. For the first time all the bits of mathematics that had been fed me through high school and my freshman year—geometry, algebra, trigonometry, and analytic geometry—began to fall into place in a larger pattern. Here, really, was another language, another means of communication, more precise than English, less emotional than music. It could arouse emotion, of course. For the conversant mind, a skillful interweaving of equations and geometry can evoke beauty, just as the reaction to a poem transcends the words it contains.

I learned to play with sequences of mathematical

themes, enjoying the unexpected symmetries that would appear. I could start, for example, with the spinning circle, generating the to-and-fro motion of a connecting rod, then jump to the differential equation of the pendulum, and modulate through its solution to the selfsame to-and-fro melody. It was a sightless, soundless beauty, but lovely nonetheless. This mathematical language also could be used as prose, to describe a motion or a shape. The description was, of course, not completely accurate, any more than a verbal description comprehensively duplicates a thought or a scene. But it was more accurate than words in some respects; it could describe symmetries and regularities more exactly and more concisely than ordinary speech.

I began to see why mathematicians enjoyed their work. Here was another medium of expression to play with and to relish. My instructors in calculus were not as inspiring as Jason Nassau had been in my freshman year. It was the subject itself that kindled my enjoyment.

Near the end of the academic year I heard that a prize would be awarded to the sophomore getting the highest grade in a special examination in physics, covering the material in the year's lectures. I usually avoid competitions and special examinations. In fact, I finally resigned as leader of a Boy Scout patrol because I never could get around to taking the examinations for First Class Scout, even though I was coaching the rest of the patrol for these tests. But here was a test I could use as an indicator of whether my choice of physics as a career was a good one, so I started to review the material covered in the course.

Education

My first action was to borrow from the Case library a physics text differing from the one used during the year. I happened to pick one that used calculus. I started to read it one Saturday afternoon and couldn't put it down until three o'clock Sunday morning, when I finally fell asleep. It brought a burst of comprehension. From the first page onward it clearly showed that calculus is the language to describe physics. The morphology, semantics, and even vocabulary of the two fields correspond completely. To think about classical physics it is necessary to think in terms of calculus, whether one is aware of it or not.

Of course, this discovery of mine wasn't new. Newton had invented calculus to describe his dynamics. But it came as a clarifying burst of understanding to me, followed by a sense of indignation at having been kept from it for so long. Here I had taken a full year's course in physics in a language other than its native tongue, calculus. It was as though one's first contact with Shakespeare had been through Japanese translations of his plays.

I spent the rest of Sunday translating my knowledge of physics back into terms of calculus, and on Monday I went in to the examination almost in a trance. All the questions seemed elementary, and the answers practically wrote themselves, but my sense of being defrauded remained. When I was called in to Professor Miller's office to be told I had won the prize, I was torn between two reactions. Of course I had won; in a sense my use of calculus had given me an unfair advantage. But why had the course in physics been organized so that I could win thus? I was the winner in an inequitable contest.

At the moment that Professor Miller smilingly handed me the envelope containing the prize, my internal struggle incapacitated me. With my meager capacity for politesse paralyzed by the mental conflict, I managed only to mutter a thank-you and to give a half-smile, which may have made me appear an overconfident, disdainful brat. But my determination to become a physicist had not weakened.

Several changes had occurred at the Radiolectric Shop while I was busy at Case. Dick Roberts, the last remaining member of the original partnership, was ready to move on. The independent telephone company in Cleveland had gone out of business in 1920, and my father had to look again for consulting jobs on telephone rate cases. There were a few such cases in the twenties, but not many, so he was often at loose ends. He took over my place at the shop when I returned to Case. Moreover, when more capital was needed, he mortgaged our home, over my mother's agonized protests. When I returned to the shop in the summer of 1924, Roberts had left and my high school friend Dick Barrett had been hired.

With the infusion of new cash, the shop began to expand again. We had obtained the selling rights for one of the first radio amplifier sets, one that was sensitive enough to pick up distant stations with only a loop antenna. In addition, we had a stock of Western Electric push-pull audio amplifiers, the first to produce sound of fair quality when used with a Western Electric horn speaker. We put the sets together into a portable unit comprising tuner, amplifier, and batteries, topped by a three-foot loop and the speaker horn. It could be carried in our car to the home of a prospective buyer for an evening demonstration.

Education

By this time broadcasting stations were operating all across the country, with big ones in Pittsburgh, Davenport, Denver, and San Francisco. We would start our demonstration by showing the prospect how to tune in to Pittsburgh. Later in the evening we would show him how to swing the loop and get Davenport and then Denver. If by midnight he could get occasional reception from San Francisco, even though the tuning was tricky and the fading was bad, he usually bought the set. The price, about five hundred dollars, should have been enough to keep us going, but goods that didn't sell always seemed to eat up profits from goods that did sell.

The newspapers were taking an interest in radio, and a reporter from the Cleveland *Commercial* asked me to contribute an article. Before long I was writing a story a week on items of interest to radio fans, sometimes technical, sometimes gossip. The *Commercial* never managed to compete with the major papers in Cleveland, but while it lasted it would publish almost anything that didn't cost them much—and they didn't pay me much. But writing a weekly article did give me some facility with words.

I continued writing for the paper after school started in September, but I had to abandon work for the Radiolectric Shop. The junior-year curriculum at Case was more specialized, more exacting, and more interesting. The course in differential equations, while touching only the surface of the subject, did increase my mastery of the mathematical language of physics. We also had lectures in optics and heat. But the real inspiration was the afternoon laboratory course.

In at the Beginnings

Only three juniors had elected the physics program: Herbert Erf, overweight and gregarious; Hank Bennington, on the basketball team and a hard worker; and I. Thus the course was, in effect, a series of private tutorials with Professor Miller. If we wanted to measure something, Miller would take us to the capacious storeroom, bring out the appropriate equipment, tell us briefly how to adjust it, and leave us with a few manuals and texts. If, after much struggle, we still couldn't get the thing to work, he would come back to the lab and, with a few magic twists, get the beam on the target or the needle in the middle of the scale. It was sheer delight to watch the jerky motions of his tiny hands produce accurate manipulations.

One time he took down from the shelf a simple demonstration interferometer. It was dusty, and the optically flat, postage-stamp-size mirrors all needed resilvering. In those days before vacuum-sputtered aluminum surfaces, we had to mix the tricky solutions that were supposed to deposit a thin film of pure silver on the glass. The job took us nearly a week. First we didn't clean the glass well enough, so that the silver was deposited only in patches. Then we didn't wait until the coating was dry, so we rubbed it off instead of polishing it. Finally we got a rather pockmarked mirror on each piece of glass, inserted all three pieces in their holders, and lit a sodium-yellow bunsen burner. Now our problem was to adjust the alignment of the three mirrors. In a correctly adjusted interferometer, light is split by the first mirror, with half going through and half being reflected at right angles. The other two mirrors reflect each beam back, to be recombined again and observed through an eyepiece. If the two beams reinforce each other, crests matching crests, then one sees light through the eyepiece; if they interfere, one sees no light; if the mirrors are tilted infinitesimally, one sees alternate bands of light and darkness. Since there are roughly forty thousand yellow sodium waves to the inch, the adjustment is extremely delicate. In fact, it was easy for us to persuade ourselves that the adjustment couldn't be made. All three mirrors had to be oriented just right to see the band-like fringes; how could we be sure two were correct when we went to adjust the third?

After two hours of twisting the tiny adjustment screws back and forth with no satisfactory result, it was time once more to call on Professor Miller. As usual, he greeted us with his indulgent frown, as if to say, "What, you can't do that simple thing?" and we all trooped after him back to the lab bench. He stooped to look through the eyepiece, he made a few quick twists on the screws, he half-smiled as he straightened up, and his eyes twinkled with modest triumph as he said, "There we are; now see if you can get the white-light fringes."

The white-light fringes were harder; here we had to make the length of the two paths exactly equal, instead of differing by an integral number of wavelengths. It took us until nine that night before we got them.

Those two years of laboratory, every afternoon, didn't convince me that I could be a very good experimental physicist. But they did give me an appreciation of the difficulties—and the importance—of accurate measurement. We measured the earth's magnetic field, and we lined up slit and prism to produce spectra. We calibrated string galvanometers, we timed swinging pendulums, and we compared weights to a fraction of a dust particle with a supersensitive balance. I came away from the course with the conviction that no beautiful set of mathematical equations can have any physical significance unless it corresponds with the experimental data. If I were to work with the mathematical language of physics, I must always remember that languages can also be used to lie.

Of course, there was mathematics in its pure form, a playing with the language itself, which I explored further under the guidance of Jason Nassau. I took his astronomy course; it was required for physicists. I also took projective geometry and vector analysis from him as an overload. These two courses were given in Nassau's office to a group of five seniors and juniors.

By the middle of the year I had somehow made a corner of Nassau's office into a personal study nook. The corner was surrounded by high bookshelves, most of them containing the works of the Cambridge and Edinburgh mathematicians of an earlier generation: Whittaker, Forsyth, Hobson, Chrystal, and the rest. Some I would leaf through and some I would try to study, but they were still frustratingly far ahead of me. Later in the year I helped in the observatory, and for a while in the spring I lived there as a student assistant. During my junior year the family finances had fallen too low for me to afford to live away from home. There were still no consulting jobs for my father, and the Radiolectric Shop still needed more capital. Additional financing would have meant relinquishing control, and my father was too independent for that. So I lived at home and travelled across town by streetcar, an hour each way. Naturally I seized every excuse to stay overnight at some friend's house or any other free place to sleep near Case, such as the observatory. Nassau must have seen my difficulties. He never let on that he did, but he managed to help when he could.

Education

My junior year also brought my introduction to research, to exploration of the new. I was invited to assist Professor Miller in analyzing the data from his ether drift observations. The ether drift experiment is famous as one of the foundations of Einstein's special theory of relativity. The first measurement of ether drift was made by Michelson in Germany in 1881. It was redone at Case in 1887, much more carefully, with the assistance of Professor E.W. Morley of neighboring Western Reserve; the work is now referred to as the Michelson-Morley experiment.

Earlier theories had assumed that light was carried by some imponderable, all-pervading medium through which the earth and sun and stars moved without friction. Thus it was expected that the velocity of this motion could be measured by comparing the velocity of light moving in line with the ether wind, or drift, with its velocity moving across the wind. Einstein postulated that no ether existed and that the velocity of light was the same in any direction, no matter what speed the measuring instrument was traveling. It sounded a little like saying that an airplane's ground speed was the same whether it had a head wind or a tail wind. Of course, Einstein also said there was no medium analogous to air, so there could be no head or tail wind for light, and he went on to point out other consequences of his theory that were equally upsetting to the oldtimers. Since that time, all of Einstein's disturbing predictions regarding special relativity have been proved correct many times over and with great accuracy. But at the time, the Michelson-Morley experiment was the first to pronounce for Einstein and against the ether. It indicated that if an ether drift existed, it was less than a fifth as large as would be expected from the known motions of the earth and sun.

Miller, who had come to Case after Michelson had gone on to Clark University and then to Chicago, assisted Morley in making further measurements, which again gave small or zero results. Later, after Einstein's theory had become so popular, Miller decided to try once more, with more sensitive equipment and with many more measurements. The measurements were fantastically difficult. Miller used an interferometer, analogous to the one we had struggled with, but with its two perpendicular light paths yards in length rather than inches. If the relative lengths of the arms and the mirrors were adjusted to produce white-light fringes and then were swung around by ninety degrees (and if the instrument was at rest with respect to a presumed ether), the fringes would not shift. But if there were an ether wind through the interferometer, the fringes would shift, as first one arm and then the other would lie crosswise to the wind. The amount of the shift would be a measure of the speed of motion through the ether.

The whole interferometer had to be swung around in the horizontal plane without altering the relative position of any part by as much as a micromillimeter. To avoid the long arms that might bend, additional mirrors were placed so that each beam reflected back and forth several times before rejoining its mate, with a multiplicity of mirrors thus replacing length of arms. Sixteen mirrors had to be adjusted instead of the three we students had struggled with. For stability, the whole device was mounted first on a heavy concrete base and later on a still more rigid steel truss, which was floated on a pool of mercury so it could be rotated slowly without jarring. The observer had to walk around the circumference of the pool, with his eye always at but not touching the eyepiece, calling off the fringe positions to a recording assistant. Since air currents and earth tremors, even though carefully shielded against, would produce spurious shifting, the sequence of readings had to be carried out over several dozen rotations of the truss, in order to average out the random fluctuations and leave what was hoped to be the measure of motion through the ether.

If an ether did exist, this motion would be a combination of the earth's motion around the sun and the sun's motion with respect to the ether. Even if, at one time during its orbit, the earth's motion cancelled that of the sun, six months later the earth would be on the other side of its orbit and the two motions would add. At none of the times they tried did Michelson and Morley find a fringe shift corresponding to as much as a third of the eighteen-miles-per-second speed of the earth around the sun. They had therefore concluded that no ether drift existed.

Michelson and Morley had set up their interferometer in the basement of the Main Building at Case, to minimize air currents and vibration. Miller wondered whether perhaps the earth dragged the ether with it to some extent. If this were so, one would not expect to read the full extent of the drift, particularly in an underground laboratory. He made arrangements to set up the well-insulated interferometer in an open shack next to the two-hundred-inch telescope on Mount Wilson in California. He hoped that the drag of the earth on the ether would be less on top of a mountain than in a cellar.

By 1924 Miller had amassed several years' worth of data, one batch recorded round the clock for several days in one month, another batch three months later, and so on throughout the year. He was ready to analyze the data, and he wanted someone to do the detailed work under his supervision. He chose me to do it. I wasn't to contribute deep thoughts to the research. Miller had worked out what was to be done, assisted by Nassau, who had become interested in the work. My job was to understand each step in the analysis and then to carry it out for the thousands of numbers constituting the data. I gladly agreed to undertake it.

The analysis took a lot of time, time I should perhaps have spent learning more mathematics. But it was a real chance to take a tiny part in either confirming or disproving special relativity. First I had to plot up the dozen or so circuits taken at a given hour on a given day in a given month. For each circuit, Miller had recorded the position of the central fringe when the eyepiece pointed straight north, again when it pointed north-northeast, then again when it pointed northeast, and so on around the compass. The fringes danced about considerably, so that when the velocities corresponding to the fringe displacements were plotted against the direction of the eyepiece, the points resembled the footprints of an inebriated beetle, staggering above and below the zero line (north). The points for the next circuit, put on the same plot, didn't correspond very well with the first, and by the time all the circuits in the run were plotted, it looked as though a gaggle of beetles had trailed across the paper, only approximately following one another.

However, for one direction of the eyepiece, more of the spots would be above the zero line than below, and for another direction more would be below than above. I would then take the Henrici analyzer, line it up, and carefully move the pointer up and down from one spot to the next, zigzagging slowly across the page. And lo, the analyzer would disregard all the fluctuations and indicate that the average of all this zigzag plot was a smooth oscillation above and below the zero line.

Of course, if the ether *was* moving past the laboratory, the chances were that this motion would not be in the plane of the interferometer, so the maximum deflection of the smooth curve and the angle it made with the north (zero) would be only a partial measure of the true magnitude and direction in space of the earth's motion through the ether. For the runs taken a few hours later, the earth would have rotated and the supposed ether drift would be coming from a different direction relative to the interferometer. Therefore, after the series of runs for a given day in a given month had each been plotted and analyzed, the resulting sequence of projected magnitudes and directions had then to be plotted against sidereal time, time with respect to the stars, not the sun. These two curves, one of projected magnitude and the other of projected direction, each against sidereal time, would also oscillate up and down, corresponding to the rotation of the laboratory in space as the earth rotated. And from each of the curves, the Henrici analyzer would finally indicate the direction of motion of the laboratory with respect to the presumed ether on that day. Likewise, from the amplitude curve, the investigator could obtain the "drift velocity" of the earth for that day. Since these values would be expected to change from month to month as the earth moved around the sun, the whole sequence of curve plotting and analysis had to be done for data from four dates spaced evenly throughout the year. It was a fair amount of work, but my enthusiasm was equally great; this was exploration of the unknown. I had occasionally to pull myself away from the work to catch up on my class assignments.

Complications showed up as soon as the first projected direction and magnitude curves were plotted for the first monthly batch of data. Of course, the measured magnitude turned out to be much smaller than eighteen miles a second, just as had been reported earlier. But Miller was assuming that the earth dragged the ether, so the full effect would not be expected near the earth. But there did seem to be some effect, and my quick plot, finished late one Friday afternoon, seemed to show that the direction of the drift, obtained from the magnitude curve, checked reasonably well with that obtained from the curve of projected directions. Furthermore, it seemed to be in the direction that astronomical observations showed that the sun was moving with respect to the stars around us.

Miller came in just as I was finishing this first analysis and, of course, was highly excited. I nervously explained that this was only the first plot; it would have to be checked and all the other batches would have to be analyzed. But he said he had waited long enough. He was going to send in an abstract for a paper to be given at the next meeting of the Physical Society. He could always modify it if the rest of the analysis required it. And he sent me home for the weekend.

Education

That night I couldn't sleep. I went over in my mind the steps in the analysis—and was brought up short by a horrid thought. Did I or did I not get the signs right in the analysis of the directional plot? I got out of bed and looked over the few calculations I had at home. It began to seem more certain that my very last step was wrong. Such an error could mean that the direction of the "drift" as obtained from the curves of projected direction would not correspond with that obtained from the amplitude curves. But I had neither data nor analyzer at home, so I couldn't tell how great the discrepancy would be.

This was the worst weekend I have ever spent. There was no one with whom I could talk. How was I to break my news to Miller? It was my mistake, but I somehow felt he should have waited at least a few days before sending in the abstract. Of course, the temptation did occur to me to say nothing; who would ever carry out the labor of checking all that work?

When the weekend was finally over, I cut classes to get back to the laboratory. Sure enough, I had made the mistake, and the directions from the two plots differed by as much as sixty degrees. Miller had not yet come in, so I took my troubles to the only friend who could help, Professor Nassau. He comforted me and sent me back to make my confession. I don't remember much of that session, except that it went off quietly. Later that day the three of us, for the first time, went over my whole analysis, verifying the discrepancy in directions, and then tried to see what could be done. Perhaps the drag also affected the direction of the drift—but then why would it differ in the two curves?

There was nothing for it but to analyze all the other

batches of runs for the other dates around the year. They all came out with similar discrepancies. I carried through the rest of the work, but I no longer believed in ether drift. The discrepancy in indicated direction, when added to the fact that the magnitude was much too small, meant to me that the fringe shifts Miller measured were caused by something else, not ether drift. What it was, I could not tell; I had never taken part in the experiments.

Miller gave his paper, but he made it a preliminary, rather than a final report. I never found out whether he finally became convinced that there was no ether. Even then I could sense the duality of his reaction to me. He might have been hasty in sending in the abstract, but I had raised his hopes and then cast them down. So I never brought the matter up and neither did he.

Years after Miller's death, Professor Robert Shankland, who had taken over the physics department at Case, re-analyzed the data and came to the conclusion that the fringe shifts were produced by temperature gradients in the laboratory shack. Probably so. A negative result is never as clear-cut and satisfying as a positive. At any rate, the ether drift experiment does not have to be done all over again; Einstein's theory of special relativity has by now amassed many other proofs of its validity, from experiments giving positive, rather than zero, results.

After such a discouragement, no further measurements were planned, so I had no chance to go West to Mount Wilson observatory, as had been hinted earlier. Instead I went to work for Nassau at the Case observatory. I made no astronomical observations—the telescopes were not set up for research then—but I would spend clear nights looking at the heavens through the twelve-inch refractor. In my senior year, Nassau suggested that for my thesis research I use the Henrici analyzer to measure the motion of the sun with respect to neighboring stars. I plotted the proper motions and radial velocities of nearby stars against their directions from the sun. I then used the analyzer to average out individual stellar motions to obtain the motion of the sun itself with respect to its neighbors. The method was quite similar to that used in the ether drift analysis. The result checked nicely with earlier measurements, made by other means, and was considered worthy of being published. The calculation was reported in a joint paper with Nassau in the *Astrophysical Journal*, my first scientific publication.

If Nassau and I had been a bit more clever, we would have noticed a lack of symmetry in the curves and thus would have discovered the "star streaming" that Oort found a few years later and used to demonstrate the rotation of the galaxy. Even so, I had contributed to knowledge, although my work was done under guidance and the result only confirmed earlier analyses. I was beginning to take a part, albeit a small one, in research; it helped me regain self-assurance and dedication.

During my junior year, the work of ether drift and my classwork kept me continuously immersed in physics. The new vistas opened up by the more intensive courses and the excitement of taking part in research filled my thoughts to the exclusion of nearly everything else. The time spent in commuting across town prevented much further exploration of the city, but I did keep up my reading. I discovered H. L. Mencken and read James Branch Cabell and Carl Van Vechten and Joseph Conrad. Having completed my course in French, I tried to read Anatole France and Rabelais in the original. It was hard going; I begrudged my slowness, the more annoying since I could read English so rapidly. The head of the foreign-language department at Case suggested that I read some old French, so I struggled through the *Chanson de Roland*. He gave me photostats of the manuscript of *Aucassin et Nicolete*, which I translated, lettered and illuminated in the appropriate style, and finally bound into a small book. But these efforts were minor distractions, to take up occasionally when I became surfeited with physics.

In my senior year, however, I swung away from my deep concentration on physics, in part in reaction to the heavy grind of the ether drift work. I lived on the east side again, which gave me two extra hours a day and put me closer to the center of Cleveland's activities. I had become a close friend of J. R. Martin, a young instructor of electronics at Case. Jack was a radio amateur, had frequented the Radiolectric Shop, and was teaching the junior course in electronic circuits. Being in the physics department, he had followed my work on the ether drift data with interest and encouragement. Through him I became acquainted with two other young faculty members, Dean Owens and Lloyd Morris.

Dean was a member of the electrical engineering department, played the cello and took part in our occasional quartet sessions. (By that time my violin playing was so bad I had to take the less difficult viola parts.) Lloyd Morris joined the physics department the same year I became a junior, having just obtained his master's degree at Wisconsin. When the ether drift work ended in the spring,

Education

he had asked me whether I would be interested in rooming with him the following winter, which would be my senior year. It was a welcome offer. After a lot of financial analysis at home, it was finally decided that the rent could be afforded, and in September 1925 I moved, with Lloyd, to a large, pleasant, second-floor room. We got on well; he was kindness itself, and I have never been very noisy, even when drunk.

First with Lloyd and then on my own I began to meet girls. I at last came to realize that those once-mysterious creatures were human beings and that it was fun to explore, with them, their thoughts and their desires. Some girls, I found, could contribute to our bibulous symposia. I soon found also that while many of them shied away from unconventional behavior, just as did most males, an intriguing minority were interested in exploration, both intellectual and tactile. Discovery that this exploration gave pleasure also to them made partnership exploration even more attractive.

It took courage to be a sexually emancipated girl in those days. The contraceptive devices then extant were contrastimulative and not very safe. Two close friends of mine had each to finance an abortion for a girl friend. But risking an unwanted pregnancy was only the final hazard. It was a big enough break with custom for a girl to take a drink or to smoke a cigarette. But once a girl *had* broken free, relationships between the sexes were often based on a degree of equality far greater than that generally found in, say, the 1950s.

My casual impression is that during the fifties, when

my children were doing their sexual dallying, the girls were expected to be completely passive; the boys made all the decisions. In the twenties, at least to my knowledge, there was an approximate equality in the relationships of those who had broken free. If the girl had rebelled at all from the rigid mores of the establishment, she took her part in the discussions and decisions as an active partner, rather than as a passive inferior. To me, this made it all the more fun. Of course, such freedom had to be concealed totally from most of the older generation. The first time my wife took a drink in front of her parents was after our honeymoon.

In 1926, it wasn't long before the field narrowed down for me to two girls and then to one. That one lived nearby and went to a college not far away. We went to concerts and to drinking parties. Gordon Williams had married, and our symposia at his apartment now had mixed attendance. As the spring advanced, so did our pleasure in each other. It is hard to be certain, but I believe we were no more inhibited in our lovemaking than is the present younger generation, although perforce we were more careful and secretive.

That spring, at Miller's suggestion, I applied for admission as a graduate student, as well as for a fellowship, at the University of Chicago, Harvard, and Princeton. My family was not quite sure why I needed more education and not certain whether it could be financed. It was time for my older sister, Louise, to enter college; she had been admitted to Oberlin. The Radiolectric Shop had finally succumbed to financial anemia, and my father's most

Education

recent consulting job might not last long enough to rejuvenate the family bank account. The recommendations of Miller and Nassau must have been potent, for I was admitted to all three places. Harvard offered me their best fellowship, $450 a year. Princeton, however, offered $700 a year plus tuition, and I quickly accepted. I had no means of distinguishing the academic quality of the three schools, but Miller had gone to Princeton. He knew Karl Compton, then head of physics graduate studies there, and thought highly of him.

The summer passed pleasantly, with some work at the observatory with Nassau and a prolonged leavetaking with my girl, who was to become a junior at her college.

3
Exploration

The Graduate College at Princeton was the embodiment of scholasticism to me; I was innocent enough to be charmed by its self-conscious medievalism. I had arrived after a long day-coach ride from Cleveland and had found my way to the office of the dean of the Graduate School. He was Andrew F. West, victor over Woodrow Wilson in a bitter war over university policy, before Wilson gave up as president of Princeton to become President of the United States.

West greeted me kindly but condescendingly. He was large all over, and his rounded forehead, with its fringe of white hair, made him look a bit like a wise old friar disguised in a modern business suit. His appearance wasn't inappropriate, as I learned later, for he was a convinced classicist who tolerated students of science as misguided misfits in the venerable halls of learning, to be weaned, if possible, from their gadgets and other nonhumanist interests. That same day he would greet Kenneth Bainbridge, newly arrived from MIT, with condolences on Ken's bad luck in never attending a real college. But the dean had known Dayton Miller as an expert flautist, so his greeting to me was a bit more forgiving, although Case, in his opinion, was probably no more a real college than MIT.

He sent me off to the Graduate College, where I collected my baggage, met my roommate, and got settled. The roommate was cheerful, self-confident Arnold

Zurcher, arrived from Cornell to study political science. We tried to find some common interests as we distributed our belongings in desks and dressers. The College, on a rise of land across the golf course from the main Princeton campus, was the symbol of West's triumph over Wilson, who had wanted to mix the graduate students with the undergraduates on the main campus. With its tall carillon tower next to the entry gate, its rambling corridors and many entryways, and its gargoyles modeled from the American scene, it was pure Ralph Adams Cram Gothic, with all the modern comforts. We donned black scholastic gowns for dinner in the great hall and heard Dean West intone a Latin grace before we sat to eat. And we seldom met an undergraduate. I was impressed.

My fellow graduate students impressed me also. I hardly noticed the scientists among the humanists and social scientists and others. Some of these were Englishmen, most of whom were on Commonwealth Fellowships that provided for a stay of two years and required extensive travel in the United States during the intermediate summer. In their first year they were usually critical and superior. After their summer exploration of the hinterland they were somewhat less strident, and by the end of the second year, some wanted to stay on this side of the Atlantic. The brain drain is not a new phenomenon. I was impressed by their accents and by their ability to hold their own in an argument, no matter what the subject. A few of them took me up, perhaps as an interesting specimen.

One of these was Lindlay Fraser, an economist whom I later visited when he was a don at Queen's College, Oxford. He was a carefree pianist, with an encyclopedic musical memory. We had a game in which he would start playing part of an opera or a symphony or concerto. I was supposed to name it, or at least identify the composer, before he had completed twenty bars. Another Englishman was the physicist Ronald Gurney, who was spectacled, dark, and silent. He startled me at first because he never said "Goodbye" or "See you later" when he departed; he just left. I got used to it, though, and found it quite enjoyable to walk beside him, for most of an hour, without either of us saying a word. We did manage to communicate when we needed to, however; he and I wrote a joint paper a year or so later.

Physics and mathematics soon claimed much of my time. I signed up for analytic dynamics under E. P. Adams, electron theory under Henry DeWolf Smyth, and mathematical physics under Dean Luther Eisenhart. (Compton was in Europe that first term.) Adams was a product of the Edinburgh school of classical physics, a large man fond of horseback riding, a bachelor, his shyness covered by a portentous formality. His lectures were clear and beautifully ordered; he never glanced at the class while he spoke. Eisenhart, in addition to being dean of the faculty, was a senior member of the mathematics department. His administrative duties evidently prevented him from preparing his lectures. He would never know where he had left off last time, but, once on the track, he would go swiftly and surely, never looking at his notes. Harry Smyth was one of the younger members of the physics faculty. He and Louis Turner lived in the Graduate College and, to me, often were friends in need. Another member of the physics staff living at the College was Joseph Morris, who became a friend and colleague.

Exploration

Three courses didn't sound like much of an academic load, but I soon found that, for the first time in my life, I had to work to keep up. It was stimulating to have to stretch my mental muscles in classwork. I was delighted with the course in dynamics, taught as I felt it should be taught, using the full power of the calculus. Dynamics is a beautiful subject, sharp, precisely rigorous, with all its interconnections clearly visible, as in the paintings of Botticelli or the symphonies of Mozart. It was fun to watch confusion disappear as we applied Lagrange's equations to one after another of the tough problems Adams would assign us. I can appreciate how shocking it must have been to those habituated to this clarity to learn that it didn't apply to atoms and nuclei.

The mathematical physics course revived my impatience with my ignorance. I sat in on other mathematics courses (we were free to visit any we could find time for), and I became acquainted with some of the mathematics students. The subjects they were excited about—set theory, topology, and number theory, for example—were of little interest to me. But I could see that analysis and higher algebra were part of the language of physics, and I worked hard trying to understand the peculiar way the mathematicians dealt with these subjects, as if they were logical games rather than representations of real phenomena. I almost decided I would have to learn mathematics by myself.

Of course there were times when I had to have a change. Uncle Will McCord, my mother's elder brother, and his wife lived in an apartment on Riverside Drive in New York, just below 125th Street, looking across to the

New Jersey Palisades. I had a key to the apartment, and a cot was always ready for me there. Uncle Will had been invalided home from the Cuban war with yellow fever and had then followed his father's profession, becoming a reporter and then an editor on the Pittsburgh *Dispatch*. Later he edited a steel trade journal and then, during the prosperous years of the twenties, came to New York as a partner in a public-relations firm. He was short, thin, active, and full of tales about politics, the theater, and the underworld of Prohibition times.

Uncle Will introduced me to parties in Greenwich Village, to dependable bootleggers, and to proprietors of speakeasies, some of whom became fashionable restaurateurs after 1933. New York became for me another world to be explored, sometimes with my uncle, sometimes alone, and sometimes with fellow students. Museums, night clubs, concerts, theaters, burlesque shows, Wall Street, Central Park, Chinatown were all new stuff for an innocent. I traveled the subway and the el for fun, to see the people and the glimpsed vistas. I learned how to transfer from one line to another without using another nickel. I soon was able to start out on Sunday afternoon, with two nickels (one for safety) and my return ticket to Princeton tucked away in an inner pocket, knowing that, wherever I found myself that evening, I could get to the Pennsylvania Station and back to Princeton in time for morning classes. Those weekends, however, were rare during the first semester. I was too busy studying.

One Saturday I was in the physics department library, getting ready for an exam, when Professor W. F. Magie, head of the physics department, marched in. With his white hair, military mustache, and ruddy cheeks, he was friendly in a blustery sort of way. He spoke to the only two of us in the library: "Mrs. Magie tells me to invite two graduate students to dinner tomorrow noon. If you two will come, I won't have to look further." We, of course, said we'd be glad to come.

I knew the other student, Jack Livingood. We had several classes in common, but I had not talked much with him. He was spectacled, a picture of the serious student. He had been an undergraduate at Princeton and knew its ways, so I asked him about various matters of protocol as we walked back to the Graduate College. He invited me up to his room, where he broke out a bottle of the local applejack, a new drink to me. We had a bit, went down to the evening meal, and then returned to his room to finish the discussion. I don't remember how we finished it, but we evidently finished the bottle as well. My roommate later told me I returned calmly to our room about midnight, reported I had had to put Livingood to bed and quietly went to bed myself. I woke up late Sunday morning with the vilest hangover I had ever endured, with the horrid prospect of sharing in a solemn social function just three hours later. I dressed and made my way up to Jack's room and found him no better off than I. Jersey applejack of Prohibition times had more unpleasant fusel oils than any liquor I had yet sampled. I swore I would stick to bootlegger's alcohol after that. Even now that experience interferes with my appreciation of good calvados.

But we still had the dinner at the Magies' to get to and to live through. We decided the only thing to do was to try to walk off our hangovers. For the two hours before we were to present ourselves, we doggedly marched around the campus. It was a chilly November day, so by the appointed hour we looked healthily blown, but every time we tried to use our brains, sand would get between the gears. The Magies cheerfully paid no attention to our lack of repartee, but they must have wondered whether all that year's crop of graduate students had so little interest in food and conversation. I never was invited again to Sunday dinner at the Magies, but from then on Jack Livingood has been a very good friend.

My introduction to other physics friends was less traumatic. Ken Bainbridge and Tom Killian lived in the suite next to the entrance gate; it was a temptation to stop in as we returned from class. They, Jack, and I all were starting graduate work at the same time and were working on the same class assignments. Another member of the group, to some extent our mentor, was Joe Morris, who was several years ahead of us. He was finishing his doctor's thesis and already teaching physics to undergraduates. Joe was round, ebullient, gregarious, and wise. His multitudinous enterprises, when recounted in his rich New Orleans drawl, became fabulous adventures. He had covered the Scopes trial in Tennessee for the *Times-Picayune*; he seemed to know the inside story about everything that happened in the South.

My reaction to the impact of all these new friends was mixed. To them the social graces seemed to come naturally, and they seemed at times to be operating by standards different from mine. I wasn't sure whether the difference was because they had lived longer in the East or whether it was because they never seemed to have to think about money. At times I would try to copy their ease of manner, but then I felt I was flying false colors. When I would try to be "myself," I sometimes felt I was just being boorish. The worst strain came when I couldn't pay my share for some joint excursion. Should I simply refuse to join in, or should I believe them when they said they didn't mind assuming the cost?

By the end of the first semester I was at least beginning to feel satisfied in regard to my studies. I was able to keep up in the toughest courses, and I felt I was beginning to understand physics. Then Karl Compton returned from Europe. He talked with all of us during the first week he was back, and I had to reorder my outlook all over again. A mutual friend, said many years later, "Karl believed only the best of everyone, and one had to oblige him by doing one's best." His gray eyes looked at a person steadily; he spoke quietly and always encouragingly. His first question to each of the new graduate students was "What research are you planning to start?" And lo, within a month we were all starting research.

I had, of course, realized that all the facts and theories I had been learning about in my classes had been discovered by someone, and I had understood—indeed hoped—that eventually I would do some discovering on my own. But to be told I should start now, even before I had learned everything that was already known, was a liberating hint. Compton's talks and his attitude suggested to me that between the lines of every published paper and text lie suggestions for further research: similar measurements on other materials, further checks of a proposed theory, extensions or modifications of existing theories, maybe even a glimpse of a new theory.

Compton's influence meant recasting my approach to

the study of physics, from one of passive absorption of existing information to one of questioning whether reported results were complete and conclusions were completely valid. I had already, unreflectively, accepted the proposition that the proper scientific attitude is questioning and exploratory; the idea of exploration had been science's attraction for me. But here was someone who suggested that I should join the exploration right then, that I could learn by exploring, not learn in order to explore later.

Compton's specialty was the electric discharge in gases, now known as plasma physics. He and Irving Langmuir of the General Electric Research Laboratory at Schenectady were the two American experts in the field. Plasma physics is an immensely complex subject, encompassing the behavior of sparks, of vacuum tubes, of the multiplicity of phenomena produced when electricity forces its way through a gas to form an aurora or the sun's corona, to cause a fluorescent tube to produce light, or, as we now hope, to react to electromagnetic fields violently enough to produce power by nuclear fusion.

To understand a plasma's multiple, often paradoxical, forms of behavior, a scientist has to combine knowledge of electromagnetism with that of statistical physics and quantum theory. Even today, plasmas are less well understood than the interiors of many nuclei. Back in the twenties, the exploration had just begun to advance, although it had been going on for nearly fifty years. Barriers to understanding were plentiful: lack of sufficiently sensitive measurement devices, lack of ability to describe complex phenomena statistically, and, above all, lack of a comprehensive theory of atomic structure. The quantum theory of the atom was just then being hammered out in Europe and would not begin to help plasma physics for several years.

Perhaps my shift from passive learning to active research was too abrupt. Certainly the research I published in the following two years had many shortcomings, and even a few errors, because I didn't yet know enough mathematics and kinetic theory. But I couldn't have forced myself to continue passive learning much longer. I agreed instinctively with Compton that I, at least, would learn to walk by trying to walk, even though I fell on my face a few times.

Compton was an experimental, not a theoretical, physicist. His usual suggestion for research was that something new be measured, so most of us started to collect experimental equipment and to put it together. I spent several months blowing glass, joining together a mercury-vapor pump, a pressure gauge, and the beginnings of a discharge tube, with which I planned to make some now-forgotten measurement. In the meantime I conferred with Compton and read Karl Darrow's compendious book *Discharge in Gases*. To me, imbued with a new urge to explore, it suggested improvements in theory rather than calling for more measurements.

Darrow had summarized the results of the experiments of Compton and others in regard to the cathode fall, the region near the negative electrode where the electrons are pulled from the gas atoms and impelled toward the positive electrode. Several empirical relationships were known, regularities in the experimental data not yet understood theoretically. I tried a new approach, putting electrostatic theory and the statistical equations for atomic ionization together with an assumption that the electric field would adjust itself so as to produce the greatest number of electrons. I thought this combination might help explain the empirical equations and enable me to relate the experimentally determined constants to the basic properties of the gas molecules. With Compton's encouragement and collaboration, a joint paper was completed and sent to the *Physical Review*.

That paper has long since passed into oblivion, buried beneath our greater knowledge about individual electrons and molecules and their behavior in a gas discharge. But I learned a lot from the task. This theoretical research was turning out to be fun, even though it also was hard work. Some of it was pure drudgery, putting numbers into a formula, multiplying and adding, looking up values in tables, and comparing the answers with the measured results. But the devising of a new theory, or even the extension of a known one, is exploration, with all the excitement and trials and false starts and effort of any exploration. It is somewhat like putting together an intricate jigsaw puzzle. Here are two or more well-tried equations, from different parts of physics, that represent disparate sets of physical phenomena. Can they be combined to explain another set of measurements of still another part of physics? Can the investigator put together a mental picture of the phenomena, which will fit the equations into a logical, harmonious pattern of the process? And, finally, will the pieces of the puzzle actually join together to produce a recognizable picture; will the combined equations, when the proper numbers are inserted and the answers

Exploration

obtained, produce values that check with the ones obtained from experiment?

Even small portions of the puzzle require this same sort of mental trial and error, before the final struggle to fit the portions into a completed whole. Perhaps an equation cannot be solved exactly. What method of approximation will work best? Does the neglect of some term in the equation correspond to the mental picture the researcher has as to what is important and what is unimportant? Concepts must be picked up and tentatively put next to one another. Some are discarded, some turned around and tried again, until all the pieces fit—or the mind gives up. The more pieces the investigator can keep in mind and the longer he can keep moving them around, the better theorist he is.

This kind of research is work of a different sort than running or digging but is just as exhausting, requiring just as much pertinacity in overcoming the body's indolence. The opposition to mental effort is not as direct as to physical effort—one doesn't get a sore brain nor run out of breath from trying to put together a theory. The opposition is more subtle. One finds oneself being deflected by the slightest excuse for distraction; one subconsciously invents reasons for doing something else—suddenly one has to go to the toilet or, if all else fails, one dozes off and loses the pattern.

On the other hand, the subconscious can be a hidden help, too. Occasionally, after a long, fruitless struggle to fit all the mental pieces together, I have given up and gone to bed, waking up to find that some or all of the pieces have been joined during sleep. What had been confusing now seems clear. It is as if I had waked up to find I had finished digging the garden while I slept, without further aches or strains.

Research is solitary work. A few years ago, the wife of a famous colleague of mine, addressing the wives of physics graduate students at MIT, said, "Your husband may often be home, at his desk. But don't deceive yourself. *He really isn't there,* and woe to you if you assume you can talk to him or ask him for help."

When a piece of work comes to an end, because of either success or exhaustion, it is time for relaxation and companionship. On weekends, that first spring at Princeton, I sometimes went to New York, often with a British friend. At other times I went to a show or concert with Uncle Will. Sometimes my weekend entertainment was no more than a bibulous evening in the rooms of one or the other of the physics group, finishing a can of alcohol one of us had brought back from the city. A favorite mix was half alcohol, half orange ice. It was rather sweet for my taste but much preferable to the local applejack. To my continuing regret we never were able to uncover a cheap, plentiful supply of wine, as had been available in little Italy in Cleveland. Once in a while we would come across some real French wine, but it was far too expensive for us to buy regularly. The greatest damage done us by Prohibition, I believe, was that it reinforced our preference for strong liquor over wine or beer.

My first final examinations came and went; Adams's exam in analytic mechanics was especially demanding, but I did well on it. And then it was June and we all dispersed.

Exploration

My fellowship was renewed, and I went home feeling I had climbed the next rung toward a career in physics. The family finances were no better, but on the basis of my record I could swing a small loan from my mother's cousin, who owned a chinaware factory in southern Ohio. As a further supplement, my friend and mentor Jason Nassau found me a month's job overhauling a telescope for Mount Union College in southern Ohio.

The telescope was a ten-inch refractor, bought many years before by some wealthy amateur who soon tired of it. He willed it to Mount Union, together with enough money to build an observatory dome for it. The telescope was in terrible shape. I wasn't to touch the box holding the objective until Nassau arrived, but the mechanical parts needed a complete overhaul. The bearings were stiff with old oil and dirt, the declination and hour-angle circles were corroded, and the scales unreadable. I spent three quiet weeks cleaning and polishing all the parts, carrying them up the stairs to the dome, mounting them piece by piece on the concrete base, then testing the instrument mechanically and attaching the clock. During the final days Nassau came down. We opened the case containing the precious objective lens, painstakingly washed off the decades' grime, mounted the objective on the upper end of the tube, and then spent the whole of a clear night adjusting the axis to true north, aligning the lenses, and setting the circles to read correctly.

The rest of the summer was spent at home, working around the house, getting reacquainted with my family, playing with some ideas for research, and, not least, picking up the threads with the girl I had known so well before I went to Princeton.

In at the Beginnings

The next academic year, 1927–28, went smoothly as far as classes went, swimmingly as regards research, but sadly in respect to Uncle Will. His wife had died the previous spring. He had aged perceptibly over the summer. He was no longer interested in his work and seemed to cheer up only when he could take me out around the town or sit in the apartment with me and reminisce. By late fall it was apparent that, like his wife, he had cancer, and by the first of the year he was bedridden. I spent as much time with him as I could. He died in March. I, as his nearest relative, did what little I could to help his close friends terminate his affairs. There was very little left after all the bills were paid; he had gone through all his savings in those last two years. I can still call to mind his gaunt face as he tried to tell me the escapades of his youth. It was my first close contact with death.

That year, in addition to my research with Compton, I took his course in the electron theory of matter and two courses given by Adams, one in statistical mechanics and one in electromagnetic theory. The latter was, to a great extent, static potential theory, with a little dynamic Maxwell theory at the end—a beautifully choreographed set of mathematical exercises, but hardly a preparation for future work in microwave or radar. That I retained little of the statistical mechanics was my own fault. The classical Gibbs treatment was presented, but I wasn't ready for it. I had to think much more about atoms, by themselves and in swarms, before I could begin to appreciate the sophistication and depth of Josiah Willard Gibbs, America's greatest home-grown scientist.

Exploration

Compton's course was down-to-earth and more nearly what I needed then. Compton gave us some Bohr quantum theory and showed us a little of the struggle that was going on as physicists were having to abandon more and more of the cherished concepts of classical physics if they were to understand the atom. Bohr and Planck and Einstein had put together a few pieces of the puzzle, but the subassemblies didn't seem to fit well with one another. There seemed to be no connection between them and the finished part of the puzzle, classical physics, to which they must eventually be joined. The puzzle evidently contained many more pieces than had been expected, and no one seemed to know what the rest of the picture would portray. But echoes from Germany hinted that a major breakthrough was in the making. Compton's presentation had all the dirt and scaffolding and confusion of new construction. And that made it all the more exciting.

I listened to several other courses, sampling Eisenhart's course in tensor calculus and Henry Norris Russell's course in astrophysics. This last course was interesting because of the lecturer as well as because of the subject matter. Physics, particularly spectroscopy, was beginning to free astronomers from the bondage of planetary orbit calculations and star-counting. It was showing them how to dig below the surface of the sun, how to estimate the chemical composition and even some of the inner structure of a star. Astrophysicists were beginning to estimate the magnitude of the energy poured out by each star, although they could not yet guess where it all came from. New things astronomical seemed to be turning up each month (as they seem to be turning up again in the 1970s).

In at the Beginnings

Russell was the very personification of his subject; he seemed to live in a state of perpetual excitement. He was a high-speed speaker, who used an expressive snort to punctuate his exposition. But his mind was still quicker than his speech, and as he came to a particularly provocative part of his lecture, his snorts would come more frequently and his high-pitched voice would come faster and faster, trying to keep up with his thoughts until the gap became too great—he would then give a particularly expressive snort and skip a whole paragraph. His speech would have caught up with his thoughts and he could go on again. It was a stimulating performance. That and the highly fluid state of the subject aroused an interest that was to surface ten years later for me.

I moved into new quarters in the Graduate College that fall. The previous spring Joe Morris had asked if I would share his suite in Pyne Tower. I had felt honored and immediately accepted. Our suite was the only one in Pyne Tower. The bedroom and bath were three flights up, and the large study above them constituted the top floor, with a magnificent view of Princeton. It was quiet there, except when the carillon in Cleveland Tower was ringing out its daily afternoon concert, when all conversation stopped until the last hum-note of the big bell ceased rattling the windows. The carillon was controlled by a music roll like that of a player piano. Once started, the carillon would run through its full repertoire. That spring, Tom Killian and Ken Bainbridge managed to purloin the keys to the tower, started the roll at three o'clock in the morning, and departed in a hurry. Joe and I were nearly lifted from our

Exploration

beds by the untimely concert and suffered through almost the whole program, until someone was located who knew how to shut the equipment off.

Joe was an ideal roommate, accommodating and always amusing. My collection of 78 rpm records supplemented his, and I put together a crude electronic reproducing system, with a massive pickup, impressive amplifier, and horn speaker, which got as much out of the records as was in them. Joe liked Wagner; he ordered seats for the Ring Cycle at the Metropolitan Opera for us and Jack Livingood. At other times we went to New York to the theater. I would contribute my knowledge of speakeasies and bootleggers, learned from Uncle Will, to keep us adequately supplied.

That year Robert van de Graaff returned from two years at Oxford to finish his doctor's degree in physics at Princeton. He was tall, athletically built, with an Alabaman's musically slow speech. He knew Joe Morris from his earlier sojourn at Princeton. Between them they seemed acquainted with all the important families in the South, and they would spend hours swapping gossip. Van spent most of his time pursuing his idea for a high-voltage machine. He set up equipment to find out how to spray electrons onto a moving paper belt and how to remove them at the other end of the loop. By the end of the year, he had built a model van de Graaff machine, with a three-foot metal sphere on top of an insulating column, inside which a motor-driven belt would haul the electrons up to charge the sphere. Impressive sparks could be drawn —when the belt was dry enough to hold the charge.

I also came to know a rather stiff young man, tall and thin, with an aristocratic profile and a Prussian manner and carriage. His full name was Ernst Carl Gerlach Melchior Stueckelberg von Breidenbach zu Breidenstein. He grew up in Basel, although as his name suggests, his father was heir to one of the minuscule princedoms that dotted Germany before Bismarck. Ernst Stueckelberg could not help his name, nor could he help his mannerisms, which had been drilled into him by his father. He once told me his father would not allow him to take off his coat or tie even when he was studying by himself in his own room. But he really wanted to become a physicist, and he had come to study under Compton. We came to be close friends, perhaps because opposites attract.

My research was progressing well. A Dutch physicist, W. Uyterhoeven, who had worked with Compton, had made some measurements after he had returned to Holland which didn't seem to agree with others he had made in Princeton. He had placed a probe electrode at various points along a glow discharge of the sort seen in neon signs. He found that the current to the probe turned out to be about twice the value predicted by a theory developed by Langmuir. He wrote, asking Compton's opinion. I was, at the time, trying to improve Langmuir's theory so as to include the effects of ionization near the probe, and Compton asked me whether my modification might explain the discrepancy. The result was a joint paper between two authors who collaborated by transatlantic mail. I met Uyterhoeven several years later, when he introduced himself to me at a Physical Society meeting.

My chief research that year, however, was taking a more ambitious turn. I wanted to generalize some of the theoretical procedures I had developed, in the hope that they might explain all of the behavior of the flow of electricity through a low-pressure gas. Perhaps Compton should have discouraged my attempting so large a task so soon, but as usual he was helpful and encouraging. I worked hard, trying to fit equations to measurements, searching for coordinating principles in solitary sessions often lasting until dawn. I doubt that I could now concentrate that hard that long. By spring I had cobbled together a theory that met Compton's approval, and I wrote it up. According to the introductory sentences of the paper, "Three general differential equations are set up which determine the average behavior of a discharge of electricity through a gas. Approximate solutions, giving the electric field and the concentration of electrons and positive ions at any distance from the cathode, are found for several ranges of value of the electric field."

I was moderately proud of the work at the time, but only a few years later I could see its superficiality. The three equations I had used left out a lot of important matters; they glossed over the essential atomicity of the processes by using average values, and they neglected the quantum phenomena involved, which weren't well understood then. The approximations used were crude ones; I had not yet learned the difficult art of tailoring neat, computable approximations to fit the problem at hand. I still had to learn how to use mathematics.

I have never been ashamed of the work, however, nor was I upset that the paper excited little comment. One does what one can at the time. A scientist is lucky if one in ten of his papers excites further work by others. Many years later I would try to console other graduate students, discouraged at the tiny contribution their thesis work would make, by persuading them that a thesis is a trial beginning, not an end, to their research. To me the fun in research lies in the doing, not in the gloating over its importance afterward.

Compton thought enough of the work to suggest that it could constitute my doctoral thesis. Implementing his suggestion posed some problems: I had not yet taken my general examination, which was supposed to be passed before the thesis was begun. Also, I had been a graduate student for only two years, and three years was supposed to be the minimum time required for a Ph.D. However, with some assistance, these obstacles could be overcome. I had to demonstrate a reading knowledge of French and German before I could take my general oral examination. I had passed the French exam the year before, but I wasn't able to make much progress in German—in fact, I didn't learn German until I went to Germany and had to hear and speak the language. But Professor Magie encouraged me to come in one Saturday and, by dint of translating half the examination text himself, he managed to get me to translate the other half. He decided I had passed; I didn't argue.

Then I put in a hectic week studying everything I hadn't studied before, for the general oral exam was to cover all of physics, not just the material I had taken in classes. I went in to the exam in a mood of fatalism. The examiners said later that I had "a nice bedside manner," but that was only because I was far beyond nervousness. They were polite. When Professor Smyth asked me the value of Planck's constant, I simply answered, "I don't know," and waited for the repercussions. But Professor Magie remarked, "I don't know either," and they went on to other topics that, providentially, I could answer. After a century or so they decided they had enough, and I made my way out the door, to be met with a welcome tumbler of orange ice and alcohol and words of encouragement from my friends. By the time Compton came out to tell me I had passed, I was in no condition even to feel relief.

Then I learned that Compton had saved my life by two more actions. He persuaded Dean West to award me the Jacobus Fellowship for the following academic year. This honor was given to the graduate student with the highest standing in the Graduate School and carried with it the princely stipend of fourten hundred dollars in addition to tuition. In addition, Compton arranged for me to have a job for the summer at the University of Michigan. Since by that time the Radiolectric Shop had gone out of business and my father was not earning much, I could not have continued graduate work if I had had to depend on family financial assistance.

In June, after a short visit home, I reported for work in Ann Arbor. The job, rather inappropriate for a "theoretiker," was to assist in a set of experiments on gaseous conduction being carried out by the University of Michigan physics department for the Detroit Edison Company. The principal investigator was leisurely, cheerful Charles Thomas, a research assistant in the department. I don't remember what the experiments were expected to prove; they had something to do with the properties of high-voltage circuit breakers. I doubt that I was much help, but my deficiencies didn't bother Charley. He assured me he didn't work more than five or six hours a day on the project, and he didn't see why I should work more than that either. This view was fine as far as I was concerned, because I wanted to explore my new surroundings.

Charley had grown up on a Michigan farm, as his Midwestern twang attested. He had a car, he liked beer, and he seemed to know all the restaurants between Detroit and Jackson that served full-strength beer or stronger drink. He also had a girl who had a friend who soon became my girl. The four of us would sample the beer or stronger drink, or go swimming in one of the many nearby lakes, or just wander off in pairs to make love.

A week or so after I arrived, the famous Michigan summer physics program began that year's sessions. The program had been started several years earlier, when Professor H. M. Randall, the head of the physics department, invited a famous pair of Dutch scientists, Samuel Goudsmit and George Uhlenbeck, to join the university faculty. At Michigan they lectured on the new quantum theory that had been proposed, piecemeal and hesitantly, by Planck and Einstein and Bohr and Sommerfeld and was subsequently being hammered into a logical whole by Kramers and Heisenberg and Schroedinger and Born.

Randall had raised enough money to put on each summer a special program to which physicists came from all over the country. There they heard some of the new theory expounded by Goudsmit and Uhlenbeck and also by others brought over for the summer from Europe. In the summer of 1928, H. A. Kramers was lecturing on quantum theory, and Paul Ehrenfest was lecturing on statistical physics.

Ehrenfest was older than the others, but he had the mental resilience to keep up with the new developments.

Exploration

He was a new experience to me; he had less self-consciousness than anyone I ever knew. If he didn't understand the simplest, most obvious statement, he would blurt out, without shame, that he didn't understand it. And I would wonder whether he knew anything. And then he would ask a further question or two that made me realize I didn't really understand it either. And after an hour's argument, never heated on his part, I would begin to think I did finally understand it ... a little. He was a bit like Socrates, if Socrates had been interested in physics.

The other three lecturers were an interesting Dutch contrast. Sam Goudsmit was short, dark, ebullient, humorously self-deprecating. George Uhlenbeck was tall, dark, slow-spoken, and gentle. These two had become famous for their work linking the multiplet structure of atomic spectra to the spin of the electron and their demonstration that this spin had half the value that classical physics would assign to it. They both had settled in Ann Arbor with their wives, each of whom matched her husband in appearance and manner. Both men's lives were to be fractured by World War II. George came from the Dutch East Indies, where the Japanese were to intern his brothers. Sam's parents lived in Amsterdam; they vanished into one of Hitler's concentration camps. But in the summer of 1928 this was in the unimagined future.

Kramers was more like my preconceived image of a Dutchman: blond, big-boned, and deliberate. He was older than the other two and held a professorship at Utrecht. Being in Michigan just for the summer and by himself, he lived in a rooming house, the same one I was staying at. We found a common interest in exploring places to eat and drink. As he explained it, he preferred to get his vitamins by drinking beer rather than by eating salads, which he called goat food. With Charley Thomas's guidance, and sometimes with the use of his car, we managed our explorations fairly well. Ernst Stueckelberg turned up halfway through the summer. He also liked beer, so the twosomes became threesomes.

My work on the gas-discharge project kept me from attending the various lecture series in the summer program, but I went to all the late-afternoon seminars. Indeed, I gave one of them, on my theory of electric discharge through gases. Under the gentle prodding of Ehrenfest, I began to wonder whether the theory explained as much as I thought it had. But hearing the encouraging comments of Goudsmit and Uhlenbeck, I began to see how it might do so after all. As I remember, those three did more of the talking than I did.

It didn't take much urging, on Ernst's and my part, to get Kramers to talk physics during our evening meals with beer. In fact, these meals turned into a series of tutorials on the new quantum mechanics. A marathon session took place during the day Kramers was getting ready to go back to Holland. Stueckelberg and I helped with the packing while Kramers lectured. Occasionally all three of us would gather around the table while Kramers wrote out a few equations, illustrating how they fitted together to explain some atomic property or other. Those casual lectures started both Ernst and me off in new research directions.

I returned to Princeton in September to room again with Joe Morris in Pyne Tower. By then Joe had his doctorate and was busy lecturing to the undergraduates and

Exploration

tending to the finances of his numerous Southern relatives in the hectic stock-market boom of 1928-29. Stueckelberg and I had laid out a whole series of calculations we wanted to attempt. We saw a chance to get in on the ground floor of research in quantum mechanics. The equations worked out by Heisenberg and Schroedinger and others seemed finally to be making a consistent pattern of theory for the atom. Newton's mechanics had worked well for planets and bricks and bullets, but it was obviously breaking down completely for events of atomic magnitude. The new theory seemed to join smoothly onto the old physics for large-scale events; it gave promise of explaining atomic physics at the other end of the scale. But to verify the theory it would be necessary to work out each of its manifold consequences, to see whether its predictions corresponded with the known facts of spectroscopy, electron-atom collisions, and molecular formation. Wherever we turned we could see chances to apply the theory, to see whether in each case it would duplicate the experimental measurements and then go further to predict the results of experiments as yet untried. In 1930 Dirac was to write, "Quantum mechanics will explain all of chemistry and most of physics." The real worth of a theory lies in its ability to predict unthought-of events. If quantum theory was indeed what we hoped, whole new worlds were now open for exploration.

But difficulties and opposition lay ahead. The new theory required a wholly new point of view. Classical physics was sharp and precise; an object was at a definite place at a definite time, and equations were expected to correspond to experiments with as much precision as one cared to take the trouble to calculate or to measure.

Quantum mechanics, on the other hand, assumed built-in uncertainties: results could not be expressed in terms of one exact number; they had to be expressed in terms of probabilities and averages. This degree of vagueness was too small to matter in the case of bricks or bullets, but for single atoms it was as large or larger than the atom itself.

The older physicists, used to the certainties of Newton's mechanics, found it harder to change over than did the younger ones who had not yet worn as deep mental ruts. I suppose Filippino Lippi would have found it difficult to emulate the work of Seurat, even if he were willing to concede that Seurat presented a valid, although different, representation of nature. To us youngsters, who had no need to strip away long-established habits of thought, the new theory had its own impressionistic beauty.

My resolve to explore this new beauty was reinforced by the arrival that fall of two new members of the physics faculty at Princeton, Edward U. Condon and H. P. (Bob) Robertson. Both had just returned from Germany, where they had spent several years absorbing the new theory after finishing their doctorates at Cal Tech. Both were round and solid, with a Western casualness overlaid with the precious vulgarity of Germany in the twenties. Condon could be distinguished by his close-cropped wire-brush hair; Robertson had a dab of mustache. They both encouraged and assisted Ernst and me to get started.

In a far-fetched sense, the work in quantum mechanics was a return to an undergraduate interest of mine. My old professor Dayton Miller's chief interest had been in

Exploration

acoustics—his ether drift experiments had been a side excursion. I had assisted him in analyzing recorded sounds before I used the Henrici analyzer for the ether drift work. I had also helped him work out the acoustic treatment of an auditorium for Cleveland's convention hall. Quantum theory emphasized the wave properties of electrons, and the wave equation of Schroedinger had close analogues with the wave equation of sound. So I took up my study of waves where I had left off at Case. This time Robertson's friendly advice helped me dig deeper. I immersed myself in series solutions, polynomial eigenfunctions, and perturbation techniques. I began to think I could tackle some of the problems Ernst and I had been contemplating.

We first tried to work out how electron waves could hold two atoms together to form a molecule. Schroedinger had found an exact solution for the standing waves of a single electron bound by a proton to form a hydrogen atom. He had shown that the allowed energies that produced its spectrum corresponded exactly to those worked out by Bohr on the basis of the earlier, partly empirical theory.

Heitler and London had just published a very approximate calculation of the interaction between two hydrogen atoms as they are brought together to form a hydrogen molecule. We thought we could learn more about this simplest of chemical reactions if we studied the system of two protons and only one electron which formed a hydrogen molecular ion. Experimental evidence indicated that the single electron could bind the two protons into a stable molecule. Of course, the single electron had to be sometimes around one proton and sometimes around the other, so the electron wave had to extend from one proton to the other. Our job was to see how the atomic wave solutions of Schroedinger, around each separate proton, could be fitted together to form a molecular wave solution around both, as the two protons were brought together. Our calculations were crude, but they clarified some questions about molecular wave functions. We felt we had made a contribution. Edward Teller, a year or so later, worked out the exact solution for the ground-state wave function and showed we had been on the right track.

In spare time between my work with Stueckelberg and my classwork (I took only two courses that term: quantum mechanics from Condon and spectroscopy from Turner), I took to playing with other solutions of the Schroedinger equation. I covered reams of paper with attempts to find the kinds of forces between two particles that would produce standing-wave solutions made up of simple functions, ones with values given in the numerical tables of that time.

One solution was well known, the harmonic oscillator. In this case, the force is elastic; it is zero when the two particles are together and increasingly attractive as they are pulled apart. The allowed energy levels are equally spaced, extending ladder-like indefinitely upward from the lowest. I investigated other kinds of forces, partly to gain proficiency in solving the equation and partly because I enjoyed playing with the wave equation. Sometimes I would start with the expression for the force and try to solve for the standing wave. At other times I would work backward, assuming the sequence of standing waves, to see what sort of force would produce it.

Exploration

One evening I found a solution for a force that was repulsive when the two particles were close together, was attractive when they were farther apart, and vanished when they were very far apart. The allowed energy levels were not equally spaced, as in the harmonic oscillator; their spacing diminished as the ladder was mounted, until finally there were no more stable states—the two particles had been pulled apart.

As I mused about this pretty exercise, it occurred to me that the force was somewhat like that between two atoms as they are brought together to form a diatomic molecule. As the atoms are brought nearer, they are more and more strongly attracted at first; then the attraction decreases to zero at the spacing of the stable molecule; if the atoms are pushed still closer, they repel each other. And then I remembered something I had read concerning the measured behavior of diatomic molecules, as they vibrated to and fro about their stable separation distance. Their vibration spectrum indicated that the energy levels were not equally spaced, as they would be if they were true harmonic oscillators. Instead, the levels moved closer together as the ladder was mounted, just as in my exercise. I realized I had stumbled on a quantum mechanical representation of a vibrating diatomic molecule; from it I could compute the interatomic force, if I measured the spectrum. It was just a nice little application of the theory, waiting for someone to come along and pick it up. It didn't take me long to finish the paper. My solution had given me the relationship between the shape of the force between the atoms and the spacing of their vibrational energy levels. Fairly simple formulas related the equilibrium distance between the atoms and the amount of work required to pull them apart, together with the spacing between the lowest two energy levels and the rate of decrease of this spacing as higher levels were reached. I then looked up the data on vibrational spectra of a dozen or so diatomic molecules and used these data to predict the energy needed to pull the constituents apart, according to the model. The correspondence between predicted values and those obtained from chemical measurements was pleasingly good.

This little result, combining my mathematical solitaire with some casually remembered facts about molecules, has, it turns out, permanently inscribed my name in all texts on quantum chemistry. The particular force field, expressed as a related potential field, is known as the Morse potential. Not a bad result for two weeks of spare time.

Condon had come back from Europe convinced that it was time to write a text on quantum mechanics in English. "How are we going to educate the next generation," he would say, being almost of the next generation himself, "when all the literature is in German?" There was so much work to do, however, that his writing went slowly. He asked me if I wanted to help. I said yes; I liked to write. The work on the book might mean fewer published papers, but I would learn quantum mechanics with the solidity one gets only by teaching it. My thesis was finished; I might as well be a coauthor. We finished *Quantum Mechanics* by spring. Condon and Morse had a long life; a paperback edition was published in the 1960s.

I began to meet more and more of the small but growing community of American physicists, a number of whom visited Princeton. John Tate came from Minnesota, a big-boned, energetic Viking, to report on his measurement of electron collisions. Robert Brode from Berkeley stayed the year at Princeton to work with Compton.

And I did some visiting myself. I gave a colloquium at Harvard on molecular vibrations, and met E. C. Kemble, who was teaching the course in quantum mechanics there and also writing a text, more ambitious than Condon's and mine. I also met a thin, high-strung postdoctoral fellow by the name of Oppenheimer, who gave me a bad case of inferiority by talking mysteriously about Dirac electrons and quaternions. I didn't know what he was talking about and his talk didn't enlighten me. Oppie always affected me that way; I never could figure out whether his sibylline declarations were just a form of one-upmanship or whether he really did see a lot more in a theory than I did. Some of both, I finally decided.

Einstein also would have a depressant effect on my ego, but for quite different reasons. (I first met him the following year.) Einstein seldom spoke unless he had some real information to communicate; he would often sit staring through me in a manner that would crumble my inner defenses, until I realized he wasn't seeing me at all—he was just thinking. In contrast, John von Neumann, who I believe was more brilliant than Oppenheimer and more versatile than Einstein, always was aware of, and friendly to, you as a person. He was interested in what people had to say and would try to make his explanations clear without appearing to be talking down to you.

I attended a meeting at the Franklin Institute in Philadelphia to hear a lecture by H. A. Lorentz, one of the last surviving great physicists of the 1900s. He fitted the role, with his white beard and a trace of Dutch gutturals in his precise and somewhat old-fashioned English. And I went to the meetings of the Physical Society, where all the ambitious young men struggled to condense their research reports into the ten minutes rigidly allocated to each paper. After seeing one unfortunate cut off by the chairman before he reached his conclusions, I practiced diligently before my first appearance, with the result that I finished in eight minutes and then spent the next hour wondering what I had left out.

Ernst and I gave a paper at the New York meeting of the Physical Society, then always held at Columbia, which raised an argument. A graduate student at Columbia had carried out a very difficult measurement of the rate of recombination of electrons and helium nuclei to form singly ionized helium atoms. He claimed to have found sharp increases in this rate whenever the relative velocity of electrons and nuclei was equal to the speed of the electron in a bound Bohr orbit. Now, orbits of electrons in an atom were relics of the earlier, semiclassical atomic theory. In quantum mechanics there is no such thing as a definite orbit. Bound electrons are standing waves, and it seemed quite unlikely that there should be any sharp peaks in the plotted curve of recombination rate against speed of approach of the electron.

Oppenheimer had carried out an approximate calculation of electron recombination a year earlier, and his curves showed no such peaks. But it was faintly possible that the approximation would overlook the narrow peaks, so Ernst and I worked out a way to calculate an exact solution for the process. It involved a great deal of algebraic manipulation and the evaluation of several contour integrals, which gave us both some intensive practice in keeping track of terms and signs and coefficients in long series. In addition, of course, we had the drudgery of substituting numbers for symbols and doing the arithmetic without the aid of present-day tables and digital computers. Our curves, when we finally obtained them, had no peaks at all; they dropped smoothly and prosaically down toward zero with increasing electron velocity.

When we reported this at the Physical Society meeting at Columbia that winter, Professor Bergen Davis, who had supervised the student's research, took up the cudgels. He first questioned the accuracy of our calculations; then, when we had assured him there was no approximation involved, he declared with some glee that the discrepancy between our results and those of the experiment proved that quantum mechanics must be a bad theory. We had no answer but to agree that if his student's results were correct, then there was something basically wrong with quantum mechanics—but we didn't believe it was the theory that was wrong.

The argument aroused the interest of Langmuir, physics' best detective since R. W. Wood. He visited the Columbia experiment and noticed that the student was both adjusting the potentiometer controlling the electrons' speed and counting the scintillations that measured the recombinations. Langmuir took over the potentiometer and asked the student to take the counts; thus the student wouldn't know the electron speed when he counted the recombinations. Sharp peaks began turning up at all sorts of different speeds. It was not that the student had necessarily fudged the results; with measurements as difficult as

In at the Beginnings

counting scintillations he could well have imagined he saw more when he expected more. Davis had been so eager to discredit a theory he disliked that he neglected to make an elementary test of experimental objectivity. The poor student had to start over on new doctoral research, and our calculations were vindicated.

The Columbia meetings were held in January, so the weather usually kept us inside Pupin Hall, which housed the physics department. The Washington meetings of the Physical Society were held in April, however—often when the cherry blossoms were out. We met at the Bureau of Standards, then spread out over an extensive and well-landscaped campus off Connecticut Avenue, just beyond the zoo. Those of us from farther north could look forward to a double spring, once at the Washington meeting and then, a few weeks later, at home. These meetings in the twenties and early thirties were like social gatherings of a few hundred friends and acquaintances. At them we could exchange gossip and ideas about our hobby and life work, physics, insulated from the world that hardly knew we or physics existed. No science reporters covered our talks—unless Einstein gave one; there were no manufacturers' and publishers' exhibits. The money we had to spend wouldn't finance an exhibit booth.

After a few meetings I could call nearly a quarter of the attendees by their first names. News traveled fast in such a close-knit clan, and consensus could easily be reached. Most of the formal intercommunication took place in January, in the corridors of Pupin or in the restaurants nearby. Although in April we might attend one or two sessions of particular interest, the rest of the time we would spend on the Bureau's extensive lawns, soaking up the spring sun and talking physics. In the evenings many of us would go down to the oyster houses along the waterfront, where Chincoteagues were fifty cents a dozen. Most of us had few funds for travel other than to these meetings. Once or twice a year we might be invited to visit another campus, and in the summers a few would make leisurely pilgrimages by car to various centers of research.

Sometime in the winter of 1929 my girl from Cleveland came to New York. We soon found we had grown apart, and after a month she returned to Cleveland. During her stay, though, I had come to know one of her Cleveland friends, Annabelle Hopkins, who was working in New York. Annabelle and I found we had so many friends in common that it seemed amazing we had never met when she was going to Western Reserve and I was going to Case. Our visits, mine to New York and hers to Princeton, became more frequent and intimate until, in the spring, we decided to marry. Both families eventually took it with good grace.

4
Fruition

      Our marriage in April 1929 was, I believe, the first to take place in the new Princeton Chapel. It wasn't an impressive affair. Our parents weren't there; Joe Morris was best man; and Joe's friend Gaylord Harnwell's wife, Mollie, attended the bride. The small party seemed lost in the high-roofed choir. We were well behaved in spite of a rather fluid lunch. After the weekend both Annabelle and I went back to work, she to her job at Brentano's Fifth Avenue Store, I to reading proof for the quantum mechanics book. By June I would have finished all the requirements for the doctoral degree and we could start living together.

    Our life was already fairly well laid out for the next two years. Princeton had asked me back as an instructor in physics for the next year, and Compton had promised to back me for an International Fellowship for the year after that. He had also suggested that, if we wanted to live in New York during the coming summer, he would ask his good friend C. J. Davisson if I could work for him for the summer at the Bell Telephone Laboratories, then on West Street in Manhattan. I liked the idea and went to see Davisson. Bell Labs was far removed from the operating companies that my father had fought against.

    That first visit confused me somewhat because, instead of seeing Davisson immediately, I found myself caught in the machinery of a large organization's personnel depart-

Fruition

ment. I first had to listen to a pleasant young man tell me what was going on at Bell Labs (much of which I knew already), asking me about my work, and listening casually to my answers. At last he listed a number of Lab scientists who should interview me; one of them, I was relieved to learn, would be Davisson. I went the rounds, taking up the time of a lot of people, and finally arrived at Davisson's office. "Davy" was short and thin—I always had the fear that anything more than a breeze would blow him away. His jerky diction, almost a whisper, enhanced his wry humor. His eyes smiled at me through pince-nez as he apologized for the confusion the personnel department had managed to produce.

He asked me what I wanted to do if I came to Bell Labs for the summer, which was the first question asked that day that I had been prepared to answer. I responded that I had become interested in the wave solutions for electrons in crystals; in other words, I thought I could work out the behavior of electron waves in the presence of a regular lattice of crystal atoms. Davisson and Lester H. Germer had been studying the reflection of an electron beam from a metallic surface, work for which they would later receive the Nobel Prize. I said I hoped their measurements would tell me whether my expected solution was a good one. Then correcting myself, I amended that to suggest that my proposed work might give Davisson ideas for further experimentation. He grinned conspiratorially and sent me back to the personnel man, who said he would let me know when he reached a decision in my case.

In due course I heard that I had a job at Bell Labs for the summer and, what was no surprise, that it would be with Davisson. I decided not to attend graduation exercises, Annabelle took time off from Brentano's, and we went back to Ohio to grace some belated wedding feasts with our respective families, to discover still more mutual friends, and to wonder again how we had never met in Cleveland.

My first meeting with the friendly swarm of Annabelle's relatives was impressive and confusing. Her father, Martin Hopkins, was one of eight brothers, sons of a Welsh ironworker, who immigrated to Johnstown, Pennsylvania and then, after the flood, moved on to Cleveland. All the sons had made names for themselves, from the oldest, Jeffrey, who was a Presbyterian minister, through William, who was city manager of Cleveland from 1924 to 1930, to Arthur, a theatrical producer in New York. Most of the brothers had large families who got together often. I found the task of remembering who was related to whom a more difficult task than remembering Maxwell's equations. The Hopkinses were all hospitable, however, and a number of them were both congenial and had congenial interests.

We returned to New York with a multitude of wedding presents and moved into a one-room apartment in Greenwich Village, on the top floor over a pleasant speakeasy. We bought a bed and some pots and knives and managed to store most of the bulky presents under the bed. Each weekday morning Annabelle would walk over to Fifth Avenue and take the bus to Brentano's, and I would pick my way through the Spanish district to West Street, next to the Hudson River. On those evenings when we didn't eat at home, we would try out some of the speakeasies that were constantly opening, as others were being raided and closed.

Fruition

When it was hot we would take the subway or el to the Battery, pay our nickels to ride the Staten Island ferry, enjoy the sea breeze, dine at one of the beer-serving restaurants in St. George, and return at dusk, enjoying the lower Manhattan skyline. We frequented Gray's cut-rate ticket agency and saw plays at reduced prices. In the spring we had seen the now-forgotten play *Wings Over Europe,* in many ways so prophetic of Oppenheimer's tragedy. That summer we saw *Street Scene, Journey's End,* and, of course, the production of Annabelle's uncle Arthur Hopkins, *Holiday,* where we went backstage to meet its star, Hope Williams. She was curtly gracious, Annabelle was mutely worshipful, I was embarrassed.

Annabelle's boss at Brentano's was Bellamy Partridge. He and his warmhearted wife, Helen, lived in an apartment over another speakeasy a few blocks from us. Bellamy was then editor of Brentano's trade magazine, *Bookchat,* and was beginning to write his series of biographical and autobiographical books. He was considerably older than we were, had grown up in upper New York State, and could keep us amused for hours with reminiscences of his country-lawyer father. He knew many of the literary and semiliterary residents of the Village; we met some of them through him and heard much about others.

My work at Bell Labs was easy and enjoyable. I sat in Davisson's office, at a desk usually occupied by Karl Darrow. Karl, for many years the secretary of the Physical Society, was away in Europe that summer. Since Davy spent much of his time in his laboratory, I was usually left alone to brood over the problem of waves in a periodic lattice and the related questions about the Mathieu equation and the Hill determinant—when I wasn't looking out the window at the Hudson River shipping. The problem I was concerned with was one whose solutions could not be expressed in terms of already tabulated functions. It would have taken me several years to compute the relevant functions by hand, and there wasn't even an adding machine in the office.

However, in another part of the rambling building there was a computing section, provided with hand-operated desk computers—the only computers there were then—and with people who could carry out any calculations they were asked to do. Davy arranged for me to call upon their services, but the arrangement was never very satisfactory. With engineering calculations, where it is known exactly what is to be computed, such collaboration would be effective, but my work was an exploration; I didn't know exactly what was required. For some reason, it was considered unseemly to talk face to face with the person doing the computations and even more improper to lean over his shoulder while he did them. I would have to put in a requisition asking the computing section to compute the functions for a series of values of the parameters I hoped would be appropriate. A week later the answers would come back, and perhaps one in ten of them would be useful. In this inefficient way I made progress, but I could see it would take me longer than the summer to finish the work I had outlined.

The most enjoyable part of my time at Bell Labs was my participation in what was then called the Three-Hours-for-Lunch Club. Lunch didn't always last three hours and the club didn't meet every weekday, but at least once a

Fruition

week Davisson, Germer, and others, such as Davy's boss, Mervin Kelly, and Walter Brattain, would explore the constantly changing universe of Village speakeasy restaurants. Since a restaurant serving wine or liquor could not advertise during prohibition, each place, when it opened, would have to serve such superlative food and wine that it soon would become famous by word of mouth. Therefore the optimal strategy for the customer was to find out about a new restaurant as quickly as possible and to eat there until it became so popular that its prices began to rise and its cuisine to deteriorate—and then to look for a still more recent opening. The game kept the club busy and well fed throughout Prohibition.

I enjoyed the company of the whole membership, but particularly that of two founding members, Davisson and Lester Germer, the other half of the Nobel Prize–winning Davisson-Germer experiment team. Germer was quieter than the others, but every now and then, after an extra cocktail, he would disclose a part of himself that seemed at complete odds with his studious persona. He was an experienced rock-climber; for him, the steeper the cliff, the better. He had been an aviator in World War I, and once in a while he would recall his experiences in dogfights over the trenches. Once he told about the time he was shot down, landing upside down in no-man's-land, to escape the following night with only a scratch from the barbed wire. He died recently of heart failure while leading a climb up a Shawangunk cliff.

The summer was soon over. Annabelle and I packed and moved down to an apartment in Princeton. I hadn't finished my work on waves in crystals, but I did finish it that winter, after a great number of further computations, done by myself on the department's desk calculating macine. The paper based on the work was a useful one. Even though the model for the crystal lattice was very simplified, the results checked not only the general pattern of the Davisson-Germer measurements, but also a number of their departures from regularity that had been harder to understand. When electronic computers became available about twenty years later, John Slater and his students were able to learn much more about electrons in metals and to develop much more realistic wave solutions.

Newcomers to the Princeton department that year included John von Neumann and Eugene Wigner, who had been in Germany. Johnny von Neumann was round, genial, helpful, and interested in everything. No one I ever knew could as quickly grasp the essence of one's halting explanations, point out the crux of the difficulty, and suggest ways to solve it. He would really listen, while simultaneously thinking through the implications of what he was hearing. It was at times discouraging to have him come out with the answer before I got through explaining the problem. He and Bob Robertson, between them, knew more salacious limericks than any pair I have ever met.

Mariette von Neumann, Johnny's wife, was energetically brilliant in her own way, always interested in people, voluable in English, German, and Hungarian, whichever happened to be handiest. Their daughter, Marina, is now a distinguished economist. Wigner, in contrast, was single, modestly quiet, and much too polite for this informal society. George Kistiakowsky added a Russian flavor to the mélange. He had been in Princeton the previous year, but working in the chemistry department with Hugh Taylor and I hadn't met him. He knew the Condons, how-

Fruition

ever, and the Kistiakowskys were part of a sociable younger faculty group that gathered in one house or another nearly every weekend. Annabelle and I occasionally participated, but we couldn't take the full schedule.

That year I had my first experience with the joys and sorrows of teaching. My assignment was to teach classical mechanics to the senior physics and mathematics majors. The subject was one I enjoyed, the class was small, and a sufficient number of students were really interested, so we could work up some good arguments. I soon encountered the gulf between understanding and ability to explain. Even with adequate preparation—two or three hours for each hour of lecture—I needed to experiment to find the best means of presentation. Rhetorical art is of no help to a teacher, however, unless he is excited about the subject and interested in the students. Restraint is needed also; Dean Eisenhart had to warn me gently that I was assigning too much homework.

Condon had organized an informal seminar for the graduate students and younger faculty which met to discuss new developments in quantum mechanics. Stueckelberg and I attended regularly, as did a new visiting fellow at Princeton that year, William Allis. Although Ken Bainbridge and Tom Killian had gone elsewhere after they had got their degrees, Jack Livingood was a fellow instructor in the department, and Joe Morris was still on hand. Francis Bitter, already interested in strong magnetic fields, took part in the seminars for a while, as did Julian Mack, a spectroscopist on leave from Wisconsin. The circle of physics friends was growing. We also got to know the Comptons more intimately. We came to love the whole family, especially Mrs. Compton.

In February I learned that my application for an International Fellowship had been approved for the next year. At Compton's suggestion, I made arrangements to work with Professor Arnold Sommerfeld at the University of Munich for the winter of 1930-31 and to go to Cambridge University for the following spring and summer. It turned out that Stueckelberg wanted to work in Germany next year, and he too decided to go to Munich. Will Allis also planned to join us there. His father, the Allis of Allis-Chalmers, had retired and was living in Menton, on the French Riviera. Will had grown up in France, had come over to MIT for his undergraduate work, did his graduate work in France, had been in Princeton for the year, and now wanted to try Germany before returning to MIT to teach physics. He was a welcome addition to our international travelling circus, with his ability to meet people and his interest in outdoor activities.

A little later I was invited to deliver the lectures on quantum mechanics at the next summer program in Michigan, so the summer before we were to leave for Europe was taken care of. Still later Compton took me aside, told me he was going to be president of MIT, and asked if I would join the MIT physics faculty when we returned from Europe. It was easy to say yes.

And so, in May, Annabelle and I stored our furniture, packed our clothes and books, and said goodbye to Princeton. We first went up to Cambridge to attend Compton's inauguration as president of MIT and to meet some of my future colleagues in the physics department there. Compton had already asked John Slater to be the new head of the physics department. He and his wife, Helen, entertained us and showed us around. John looked more like a gangly undergraduate than a full professor, but he was already a veteran of the new physics. He had done his graduate work at Harvard, had spent 1923–24 in Copenhagen and Cambridge, England, had joined the Harvard physics faculty, and then had returned to Europe for a year before he took up the job of building a new physics department at the Institute.

Slater was full of ideas about the new department. MIT had been primarily an engineering school, with the science departments regarded as service departments. Compton's plan was to emphasize basic sciences, and he wanted the physics department to be the best in the country. A few of the older faculty were staying on: Julius Stratton in electromagnetic theory, Manuel Vallarta in relativity and cosmic rays, Bertram Warren in X-rays. More than half the department were to be new. Some of these I had met already, Will Allis and Bob van de Graaff, for example.

We discussed what I would teach when I got back. Everyone in the department was to take part in the redesigned course in beginning physics, required of all freshmen and sophomores. Someone was needed to teach a course in acoustics to the physicists and electrical engineers. It was suggested that I might teach the graduate course in quantum mechanics.

I shared Slater's enthusiasms and hopes. At that time there were fewer than a half-dozen institutions in the country where students could learn the new physics. My own experience at Case showed how inadequate the curriculum was at most schools, both graduate and undergraduate. In Europe, youngsters in their early twenties were making major research contributions; here I was already twenty-seven and I still felt inadequately trained in mathematics and classical theoretical physics. Why should the Europeans have all the joy of discovering the new physics? Why couldn't we so teach undergraduate courses in physics that we would not have to teach the material all over again correctly in graduate school? Surely we could educate physicists thoroughly and quickly enough so they could spend their twenties, their most creative years, doing research rather than only studying.

These aspirations went against established custom. It had always been assumed that pure science could be learned only in Europe. Even in the twenties no one felt he was a research physicist until he had studied abroad; Slater and Condon and the others had done so, and I was on my way to do likewise. It wasn't that we foresaw the urgent practical need for physicists that was to come a decade later. We just felt it was time that this country was able to train its own scientists, even in such an obscure and unfashionable subject as physics. Both Slater and I felt it would be a worthwhile contribution for MIT to turn out each year six to ten well-trained physicists with doctoral degrees, and we were optimistic about finding jobs for that many. I left MIT that May in a mood of dedication.

Ann Arbor that summer looked much the same as it had two summers earlier. This time, however, my status had changed from that of a lowly research assistant to that of an invited lecturer, a coauthor of the only text on quantum mechanics then published in the United States. In addition to the Goudsmits and the Uhlenbecks, whom I had met during the earlier summer, the Fermis were there. It was their first visit to the States; they added their Italian

Fruition

flavor to the polyglot chatter at the teas and cocktail gatherings. Laura Fermi was serious, rather shy, but interested in everybody. Enrico was rather more vocal and was interested in everything.

Fermi and von Neumann were the two, of my generation, who from the first impressed me with their superlative intellectual abilities. Johnny seemed more intuitive; but Enrico seemed more logical. By asking the right questions he would gradually lead his hearer to put the pieces together in the right order. My reaction to both of them was not one of despair or of envy, but of pleasure in watching a matchless performance. With these two I had no urge to try to compete, only a delighted appreciation of excellence.

I sat in on Fermi's lectures, which came after my course on quantum mechanics. His course was on quantum electrodynamics, and he reported on the progress being made in quantizing the electromagnetic field. This highly mathematical subject was to develop into quantum field theory, the abstruse foundation for the theory of nuclear, as well as atomic, forces.

To me it was as entrancing to watch and listen to Fermi's masterly development and exposition of this evolving theory as it was to watch and listen to Pablo Casals in his prime. Casals always seemed to me to be above the details of fingers, strings, and bowing. His facial expression, during a performance, appeared divorced from the complex and exquisite things his body was doing to the cello; he seemed gently pleased and faintly proud that it was performing so well just then. Fermi seemed also to be above the details of the theory he was explaining; he saw beyond the symbolism to its basic structure and harmony, but he also had each note and semiquaver well in hand.

The climax of Fermi's lecture series came with a formula that stretched twelve feet across the blackboard, an extended counterpoint including all the electromagnetic waves that could be sent from one electron to another and that, eventually, through one variation after another, reduced to the simple motif of Coulomb's law of classical electrostatics. The bell rang when Fermi was only halfway across the twelve-foot band of symbols. The next afternoon when we came back, he had reconstructed the formula exactly as it had been when the bell rang. He started up, just where he had left off, and carried through to the final coda, as though the intervening twenty-two hours had not existed. Only a complete master could have made the theory seem so simple and inevitable.

We also made other fast friends who were more nearly at our own stage of education and ability. Many of the graduate students and junior faculty members attending the University of Michigan summer sessions were to be prominent physicists later. We came to know Bob Bacher and his wife, Jean, a handsome pair of Michiganders. Bob was finishing his doctoral thesis under Goudsmit, and Jean was taking courses in social science. They knew the countryside, had ideas about picnics, and could locate good lakes for swimming, if only transportation could be arranged. So at a Detroit used-car lot we acquired our first car, a Model T, for the sum of $25. Something was wrong with the battery, so it was desirable to park the car on a slope, but, once started, the sturdy old engine kept turning over as long as the headlights were not turned on full; evidently it was too much for the battery to attend to the

Fruition

headlights and to the ignition at the same time. If we had to drive at night we used the parking lights or, if there was a full moon and we were in the country, no lights at all. The price of the car included its license plates. Insurance was an unnecessary luxury. Until recently, I still had my Michigan driver's license, good in perpetuity, obtained without examination, on payment of one dollar.

Between lectures and discussion groups, parties and picnics, the summer went too fast. By the middle of August we had to begin the sequence of steps toward Europe. A stop in Cleveland and then to New York, where we boarded the *Saturnia,* bound for Naples.

The previous spring I had obtained approval of my fellowship itinerary from the administrators of the Rockefeller fund that financed the International Fellowships. I would have liked to visit at least five centers of the new physics: Cambridge, Copenhagen, Göttingen, Leipzig, and Munich. The Rockefeller people advised against too many short visits; they preferred a full year's stay at one place but would allow a half year each at two centers. I had been interested in the quantum properties of metals, and Sommerfeld, at Munich, had been pioneering in this field, so we were to go there for the fall and winter. I chose Cambridge for the spring and summer for less specific reasons. Neville Mott was there, and he had been doing work on the theory of metals. Also, he and Harrie Massey had been working on the collision of electrons and atoms, a project similar to the work Stueckelberg and I had done. But I also wanted to visit England to meet the many famous physicists then at Cambridge: Rutherford, Kapitza, Dirac, and the rest.

It is hard to remember all the stresses of our first immersion in a foreign culture. I had never before been where I could not understand all that was said and where I would be only partially understood. Reactions and preoccupations were different, and the differences were of an order of magnitude greater than I had experienced when moving from Ohio to the East Coast. Since half the boarders in our pension in Munich were American, most of the conversation at the dinner table was in English, so the difficulties were not just linguistic.

Frau Molsen ran the pension in a pleasantly efficient way, but Herr Molsen was a new type to me, although I later met many others like him. He had been a fairly well-to-do governmental official who had lost his job and most of his money in the inflation and political turbulence that had followed World War I. He never forgot it. Most of the time he would eat quickly and silently, except for a few remarks in German to a select few of the Germans present. But once in a while some comment would make him begin talking, in his high-pitched, palatalized English, about the sufferings of Germany and the necessity of overturning the Versailles Treaty. Later, one would have called him a proto-Nazi.

We began to make acquaintances, first among the Americans at the pension, some of whom had been there all summer. They showed us around Munich and tried to get us to go hear a man called Hitler, who raved about pan-Germanism and anti-Semitism to a coterie of fanatics. I checked in at the university, but the fall term had not yet begun so no one was around. My first quarter's check came from the fellowship fund. I struggled through the intricacies of opening an account at the Deutsche Bank and impatiently learned the correct order for joining the three successive queues in which one had to wait while cashing a check.

Will Allis arrived, and soon after that Ernst Stueckelberg. Professor Sommerfeld got back, and I could turn to him for advice on living arrangements as well as plans for study and research. (We would have to look for an apartment large enough to accommodate us and the baby we were expecting in February.) Sommerfeld was short and heavy-set, with a face that was a gentler, kindlier replica of Hindenburg's. He was a welcome contrast to the usual Herr Professor. He once told me he enjoyed lecturing in the United States where he could relax with the students in a way that custom would not allow in Germany. With his help we got a list of apartments to rent. In one of them, the coal for the stoves was in the bathtub. In another, the living room was crammed with fragile bric-a-brac, covering every table and shelf, which was not to be disturbed. Finally we came on a unique exception, simply furnished, clean, and only a short distance from the university. The rent was more than we could afford, but Stueckelberg offered to rent the spare bedroom from us. Our new landlady's maid had a sister, Rosa, who agreed to be our maid. We moved in and began learning how to run a household in Germany.

There was no central heating; each room was warmed by a coal-burning porcelain stove, towering in a corner. There was no refrigerator; Rosa went out each morning to buy the exact amount of eggs, butter, meat, and other perishables needed for the day. The only supplies delivered were the newspaper and the beer, both of which arrived fresh at our doorstep each morning. Rosa got up at six, fired all the stoves, cooked breakfast, went out to buy the daily victuals, and cooked lunch and dinner; in between these chores she scrubbed, dusted and tidied every part of the apartment. She spent the evening washing, ironing, and repairing our clothes and polishing our shoes. She thought we were wonderful to give her Sunday mornings and Thursday evenings off; our landlady complained mildly of our being too lenient with the help. We still exchange yearly letters with Rosa.

When the fall term began, I sat in on as many lectures as I could, in part to learn the language. Sommerfeld's lectures were doubly rewarding, for the German and for the physics. Vocabulary is just the start of mastering a language; one also has to hear the timbre of the words and feel the pattern of well-spoken sentences. I began to feel that German wasn't so impossible after all. As regards the physics, Sommerfeld was a master of the application of mathematical analysis to the classical theory of fields. It was what I needed; I drank it in.

The research didn't get started quite so easily. The recent work of Sommerfeld on electrons in metals involved a great deal of quantum statistics, and I would have to do some studying to become fluent in its methodology. However, Will Allis was interested in the behavior of electrons in gases, a subject I had delved into at Princeton under Compton. Now, with the advent of quantum mechanics, some of the elements of the complex phenomenon could be studied realistically.

Fruition

For example, it should be possible to calculate in detail what happens when an electron strikes an atom. Sometimes the electron gives up some of its energy of motion, raising the atom to a higher quantum state. If the electron is going slowly, however, all it can do is to bounce off the atom. So if a stream of electrons is sent slowly through a gas, a fraction of them are scattered away from their original direction. The amount of this scattering determines many of the characteristics of the flow of electricity through that particular gas.

The quantity measuring the total amount of scattering is called the cross section for scattering of an electron from the particular atom of the gas. It can be measured experimentally (although with great difficulty). I hoped that with the new theory it would be possible to calculate it theoretically. If the calculations could be shown to agree with the experimental measurements in a few cases, it might turn out to be easier to calculate other cross sections rather than to measure them.

A few years previously, Ramsauer had measured the cross sections for slow electron scattering from a number of monatomic gases and had discovered a peculiar effect with some of the atoms. Usually, as the electrons are sent through the gas more and more slowly, the scattering becomes greater and greater; as the slower particles are more likely to be deflected. Ramsauer found, however, that for neon and argon atoms the scattering below a certain electron speed drastically diminished; these gases were practically transparent to very slow electrons.

Here was a challenge. Classical mechanics, expecting electrons and atoms to be elastic balls, could not explain how an atom could appear to be a small Ping-Pong ball to a slow electron but a large bowling ball to one going three times faster. Perhaps the new mechanics, which considered electrons to be also waves, could uncover some resonant reinforcements and cancellations that might explain the effect.

In some respects, the job was analogous to the calculation of the scattering of electromagnetic waves—light or radar—from fog droplets. Sommerfeld had been a leader in the analysis of these waves; he helped Will and me get started. The interaction of the incident electron and the atom's interior was much more complex than that between light and a drop of liquid, however. The atom contained other electrons, all moving in bound orbits, and each of them could affect the incoming electron. Of course, the energy of binding of the atomic electrons was much greater than that of a slow intruder; perhaps their effects blurred together to produce a smoothed-out average.

So we were back at the favorite game of the theoretical physicist: how to devise a plan of calculation simple enough to carry through and yet complex enough to take all the important effects into account. As with a poem or a novel or a quartet, what to leave out is as important as what to include. And as with any of these, we couldn't gauge our success until the work was nearly finished.

In the 1970s, of course, we do not have to try so hard to simplify the model; we now can keep many of the complexities and let the high-speed computer do all the work—or quite a bit of it. Our two-months' computations could now be done in a few minutes (after we had sweated through a week or so of hard work, programming the machine instructions). I'm old-fashioned enough to feel that the computer removes a lot of the thrill along with much of the drudgery. I also am sure that now we do not end up knowing as much about the essentials of the phenomenon that the computations represent. Researchers today are not forced to find out what is important and what is not, as we were in the fall of 1930. We reported our results in one of the German physics journals, in an article written in German that required the major editorial assistance of Sommerfeld and Stueckelberg. Our results checked surprisingly well with experiment.

During the fall a number of famous physicists visited Sommerfeld's institute. William Bragg and his wife roomed near us for several months. Mrs. Bragg was pregnant, and she and Annabelle became fast friends. Bragg, the first Nobel laureate I had met, soon dispelled my awe by his quiet friendliness. We seldom talked physics, for his work was in X-rays and he was not up on quantum theory. But he helped introduce me to the world of art and music in Munich. He knew the best bookstores. We went with him to performances of the Ring.

Somewhat later, Linus Pauling arrived with his wife, Helen, who was also pregnant. His Nobel Prize was in the future, but he already was famous in the physics fraternity for his work on the quantum theory of molecular structure. He was full of energy and optimism and enjoyed rambling through the older parts of Munich and visiting and palaces and castles of the Wittelsbachs, past rulers of Bavaria.

Toward the end of February, Annabelle went to the hospital, where our son Conrad was born. When I registered Conrad's birth at the local police station, the official form required the names and the birthplace of Conrad's parents and grandparents. Every birthplace was different, a matter of much puzzlement to the German policeman. How had all these people managed to meet each other?

We were often struck by differences of this sort in the way things were perceived or done. Once, early in our stay, four of us got on a streetcar to go downtown. One of us reversed one of the seatbacks, so we could all face one another. Immediately the conductor appeared, motioned us out of our seats, and, with smiles and explanations, replaced the disarranged seatback. His assumption was that we didn't know the rules or we would not have introduced disorder. After all, a tram should be neat and symmetric, with all of its seats uniformly facing forward. People in Germany seemed to like rules, to enjoy a sort of regimentation that set my teeth on edge.

Once in a while, on Sundays, we would be held up by columns of marching men, often going through intricate parade-ground maneuvers. Each of them wore a brown shirt as a uniform. We were told they were some sort of organization connected with that crazy fellow Hitler. Of course, we were told, the upper classes looked down on all that silliness.

By December we had made some acquaintances among the "proper Müncheners," through the university, through our landlady (who was the divorced wife of a famous art critic), or through Ernst Stueckelberg, whose father had numerous friends in Munich. Nearly all of them were sure "the people" would soon tire of Hitler's ideas; no one of importance was backing him. However, as we came to know them, these cultured people also unsettled us. Occasionally their casual comments would reveal attitudes toward people and toward the world which were unexpectedly alien to ours. People we thought we understood would suddenly turn out to be motivated in ways we could not understand.

The older ones, of course, had not got over their defeat

Fruition

in World War I; occasionally we caught glimpses of virulent hatred of all things French. But the younger intellectuals were affected by the future, not the past. We had left the States before the effects of the stock-market crash were apparent, but the depression had started earlier in Europe. Young men just graduating from the university could see no sure career ahead of them. Even in our secluded field of physics we could witness the social and economic pressures.

The only secure position in a German university, the only one commanding a comfortable salary and public esteem, was that of a professor. But in Germany there was only one professor in each department; all others on the staff were privatdocents, unsalaried assistants remunerated only by student fees. At first I was both surprised and pleased at the awe I evoked when I showed my passport identifying me as "Assistant Professor of Physics." Disregarded was the "assistant," a professor was a Professor. I was thereupon deferred to as a member of a very small, highly influential elite.

German universities were then turning out many more Ph.D's in physics than there were professorships available, and industrial jobs were not generally open to university graduates. The prospect before most of my student friends at the Institute of Theoretical Physics was to slave as a privatdocent for a dozen or more years or to waste an excellent education entirely by taking some unrelated governmental job (as Einstein did before his brilliant papers gained him fame) or to emigrate.

These bleak prospects bred tension, bitterness, and extremes of competition. Students were chary of discussing their research until it was completed and published. Anyone frowned on by the Herr Professor became a complete outcast. To a foreigner, of course, everyone was pleasant and friendly. Only with time did I come to see these undercurrents. In retrospect I can understand the hell that resulted when the love of regimentation, the bitterness of frustrated hopes, and Hitler's crazy ideas all came together.

Will Allis and I traveled some during the early spring. We visited Leipzig, where Werner Heisenberg was training the next generation of quantum physicists. Of the group we met there then, Felix Bloch, Edward Teller, and others, only Heisenberg was obviously Aryan enough to survive the future Nazi regime. The others later found their home and gained their fame in America. But in 1930 they were in Leipzig, working on the quantum theory of metals and molecules. I wondered whether my choice of Munich had been the right one.

In the spring, we left as planned for England, accompanied by Allis and Stueckelberg and with the baby traveling in a laundry basket. In Cambridge, we all lived together at a boarding house where we and our guests were the only residents for all of our stay.

It was easier to get acquainted with the physicists in Cambridge than it had been in Munich. The language barrier was *almost* nonexistent, and the atmosphere was much more relaxed. The Rutherford era at the Cavendish Laboratory had not yet ended, and a galaxy of great and soon-to-be-great physicists were there. They were divided, like Chaka's Zulus, into age groups, with Rutherford on top by himself. Next came the well established: Neville Mott, to whom I reported; P.M.S. Blackett, who was investigating cosmic rays then and was to win a Nobel Prize for his work; the Russian Kapitza; the solitary Dirac; John Cockroft, later director of the British AEC; and a few

Fruition

others. They held weekly discussion sessions, to which I was occasionally invited.

But I spent more of my time with the younger group, particularly with H. S. W. Massey, now Sir Harrie Massey. He and his wife were from Australia—another accent to get used to—and he also was interested in the theory of collisions between electrons and atoms. He was interested in the results of the work Allis and I had done on slow-electron collisions, but his major interest was in higher-energy collisions, when the incident electron could jostle the atom into a higher quantum state or even knock loose an atomic electron. High-energy collisions were a much more difficult problem, one that was not to be solved, even approximately, until big computers became available. Stueckelberg and I had worked on an aspect of the problem in Munich but had not succeeded very well. Massey, under Mott's supervision, was working hard on a tangle of equations and numerical calculations. He was concerned about the relative importance of the exchange effect, the chance that the incoming electron trades places with an atomic one. As this could happen even with slow electrons, Allis and I had discussed its importance, but had finally decided to neglect it—partly because we couldn't figure out how to calculate the effect. We hoped that the fact that our calculations, which had omitted exchange, had managed to correspond so well with experiment could be taken to prove that exchange effects were small at low energies. Massey did not agree.

The summer was not all work. We learned the specialized sport of punting on the Cam, at the price of a few highly comic spills. We went to Blackett's parties, and I attended some of Rutherford's high teas, where I learned to keep quiet during Sir Ernest's pungent comments about America and Americans. We also borrowed Blackett's car and drove to East Anglia.

My father, reminding me that our ancestor Samuel Mors (they spelled it that way then), had come from England to help found Dedham, Massachusetts, wrote to ask if I could find out whether he had come from Dedham, England. I wrote to the vicar of the church at Dedham, Essex, and learned that indeed there had been Morses in Dedham and that the family had built a church in Stratford St. Mary, nearby.

We drove to Stratford St. Mary, past the scenes of many Constables and through the originals of many Massachusetts towns: Haverhill, Ipswich, Medford, and Sudbury, as well as Dedham. In the little church there, its flint walls showing few signs of age, was a band of Tudor self-advertising spread along the outer side wall, just under the windows, asking us in Latin to pray for the souls of Thomas Mors and his wife, Margaret, 1499, and in English to pray for those of Edward Mors, Alys, his wife, "and all crysten sowlys, 1530." We heard from the vicar that these men, father and son, had been wealthy wool merchants. I later found that the ability to trace one's ancestors back fifteen generations, to a pair who built a church in the time of the first Tudor king can help one calmly to face the condescension of a Louisburg Square Bostonian.

Allis's father came up from his home in Menton and shamed us by getting up at dawn and exploring all of Cambridge despite his eighty years. Then Julius Stratton came over from Boston. He had been a close partner of

Fruition

Allis during their undergraduate years at MIT; they had spent summers together doing such romantic things as taking a boat down the Yukon and hiking the Alps. A bicycle trip through Wales seemed appropriate for this summer; Mott agreed to make it a foursome. We took the train to Oxford and started west, through the Cotswolds to Tewksbury and on to Clun, on the Welsh border. Here Allis sprained his knee jumping over a two-foot brooklet—after climbing a ruined castle tower. He and Stratton returned to Cambridge by train while Mott and I ground through the Welsh hills to Machynlleth. We climbed Cader Idris, circled back to Shrewsbury, and entrained to Cambridge, where Allis was better but still using a cane.

And then it was time to get ready to return. It had been a successful year. I had my name, together with that of either Allis or Stueckelberg, on five papers. I had met nearly half of the important theoretical physicists in Europe and had learned what they were working on. It was time for me to join with Slater, Stratton, and Allis in building a strong physics department at MIT. In one sense I had completed my preparatory years; I could now begin to help train the next generation of physicists. In another sense, my own most fruitful years of research were nearly over; I would soon have to resign myself to watching younger men make the bright new discoveries.

5
Consolidation

Looking back, I can see that the first three or four years at MIT set the style for the rest of my life, although I didn't recognize it at the time. During the latter part of my graduate work and during the year abroad, I had been free to concentrate on research and, through the research, on learning more and more about some particular part of physics. Now I was a member of a faculty. I also was a father. I would have to divide my time and my energies among at least four different kinds of activities: research, teaching, administrative duties, and, not least, my family. No matter how I divided my time, each activity would have to be shortchanged to some extent. Which was to have the highest priority?

Having watched the careers of several Nobel Prize winners, it is obvious to me now that if I were to have become a truly distinguished expert in some branch of physics, I would have had to concentrate almost entirely on research, cutting the time spent on the other three activities to a minimum. This could have been done. Compton's plans for MIT allowed for a small fraction of the faculty to concentrate on research.

Somehow I couldn't go that way. My urge to taste everything kept me interested in family, in students, and in my new colleagues as much as in new research. I simply could not keep pointed in only one direction. I could not have concentrated for weeks and months on one piece of one problem, as Einstein and the other greats in physics have done. I have always learned new subjects quickly; perhaps breadth rather than depth was best for me. In any case, my choice was not conscious. I just found myself following a variety of interesting leads. Perhaps the most effective life results from following one's interest, wherever it leads, rather than by assuming some rigid set of obligations, self-imposed or set by others. I was lucky, of course, in being born at the right time. A generation earlier I would not have found a career that gave me almost complete freedom to do what I wanted to do, while at the same time paying me enough to live comfortably.

During the next four years, then, although I turned out one long paper and several short ones on electron collisions, I also spent a lot of time becoming acquainted with MIT faculty members and students. I already knew Karl Compton, our new president, and John Slater, the new head of the physics department, and Will Allis and Julius Stratton and Bob van de Graaff, who were now part of the physics faculty. Another physicist, Manuel Vallarta, was a short, thin, studious, Mexican-born expert on cosmic rays, who sported a high-powered Stutz roadster. Other physicists were George Harrison, who had come to MIT the previous year to build up a spectroscopy laboratory; Bert Warren, the X-ray specialist; Wayne Nottingham, an electronics expert; and N. H. Frank, who was working with Slater to reorganize the undergraduate courses in physics.

Because most of MIT was then housed in one large building, it was easy to meet people in other departments. The electrical engineering department was just around the corner from my office. It had split off from the physics department early in the century and was the most scientifically oriented of the engineering departments. Both Stratton and Vallarta had been undergraduates in electrical engineering at a time when the physics department had been but a service department, teaching undergraduates but supporting little research. In the thirties the electrical engineering department was in the process of changing its major interest from electric power to the expanding field that is now called communications, including the applications of electronics to acoustics and to computation. Radar was then unknown and the transistor was not yet even a fancy in the mind of one of our graduate students. But much was being done with the vacuum tube in revolutionizing communications and in enhancing the accuracy and sensitivity of measurement of tiny sounds or vibrations or currents.

I soon came to know Edward Bowles, who was busy teaching about vacuum tubes and how they were transforming telephone communications. He was already consulting for several of the electronics companies that were sprouting all around Cambridge, some to damp off, others to grow to huge conglomerates. Another colleague was shy, soft-spoken Richard Fay. His two penchants were sailing and acoustics. He had been navigator on *Yankee,* one of the America's Cup defenders, and his rare figures of speech were nautical, a favorite being "If the rope's too short, it's easy to fix; if it's too long, there's hell to pay." He had been working with the Submarine Signal Company on the beginnings of sonar, combining his two interests of sea and sound. The new electronics was transforming the science of acoustics from the ear-bound empirics I had learned under Dayton Miller to a science in which frequencies could be made to order and could be precisely measured electrically. I enjoyed watching Dick Fay carry out his simple yet elegant acoustical experiments. It rearoused my interest in the science.

Another stimulating electrical engineering acquaintance was Vannevar Bush. His office was just down the hall from mine that first year, and I found it easy to drop by. He would be leaning back in his chair, his feet on his desk, interspersing puffs of smoke from his eternal pipe with bits of dry humor or laconic wisdom, spoken in his Yankee twang. At that time he was busy improving his differential analyzer, the first operable machine capable of solving differential equations. It was a fearsome thing of shafts, gears, strings, and wheels rolling on disks, but it worked and it foreshadowed a host of fantastically more capable computers. I wondered aloud if it could be used to improve the calculations that Allis and I had so laboriously carried out on the scattering of slow electrons from atoms. Perhaps it could settle some of the arguments we had had with Massey, regarding the effects of electron exchange. Bush said, "Go ahead."

That computer was programmed with screwdrivers and wrenches to connect its shafts to the disks and wheels and to the arms that moved the pen that drew the graph that was the result of the calculation. As with any computer, then and now, it took several tries before the machine calculated what was wanted; sometimes the answer went off scale and sometimes we found we had connected the wrong shafts. We finally got our answers, ones interesting enough to publish, although not the complete answer to the original question—that had to wait for more accurate and more powerful computers.

Manuel Vallarta also used the differential analyzer. He

had been interested in cosmic rays, those high-speed particles from outer space that crash through our atmosphere, producing the radioactive carbon that enables us to date archeological finds and also producing in us some of the mutations without which evolution could not work. Measurements of cosmic rays were then being made all over the world, to try to get a clue as to their source. Vallarta had a hunch that most of the particles were electrically charged, that therefore the magnetic field of the earth would deflect them, and that then their direction of motion when they hit would differ greatly from their original direction. He became convinced that this deflection would have to be worked out before one could deduce the origins and constitution of the cosmic rays from the measurements.

Manuel worked on the problem with his good friend, Abbé Georges Lemaître, a professor at the University of Louvaine but a frequent visitor at MIT. They used the differential analyzer to compute the orbits of high-speed, charged particles in the earth's magnetic field and showed that there was, indeed, a large effect. Only along the two polar directions would the particles be undeflected; those attempting to hit near the equator would never reach the earth at all unless they possessed extremely high veolocities. Vallarta and Lemaître's work made it possible to understand the major peculiarities of the cosmic-ray measurements. Since the advent of space exploration, of course, the importance of the effects of the magnetic fields of the planets, of the sun, and of the galaxy have often been demonstrated. Many calculations of the effects have been done since, but the work of Vallarta and Lemaître began it. Allis and I often ran into them as they were finishing

Consolidation

their runs, or as we were finishing ours. They made an interesting pair, Manuel so small and so precise in manner, and the abbé so large and ebullient.

In the meantime I was also getting acquainted with the students. The physics department had a large service load, since every undergraduate had to take two years of physics and many departments required their students to take still other, more advanced physics courses. Every member of our department took part in this service task. It was rewarding; we got acquainted with all the undergraduate students at the Institute. And when one learns how to explain such things as Coriolis force to freshmen, one has begun to understand classical mechanics. A gleam of interested comprehension appearing in a student's eye can repay hours of preparation.

One of the service courses offered by the physics department was acoustics, required of the electrical engineering juniors and an elective for the physicists and mechanical engineers. Because of my work with Dayton Miller I got the assignment and rather welcomed it. I had not thought much about acoustics since my days at Case. I now found that much of what I had since learned shed new light on my ideas about acoustics. The new quantum theory dealt with electron waves, and the analytic techniques worked out to deal with these waves, I began to realize, could be used to learn new things about sound waves. In addition, my talks with Dick Fay showed me that many phenomena that theretofore could only be guessed at could now be measured, so new theoretical advances could be verified by experiment.

The new acoustic course was the first one I developed by myself; it set a pattern I have followed a good many times since then. I made it a hard course, for I wanted it to be a challenge to the brightest in the class. I still feel this approach is important, especially at a place like MIT. New advances in science, as in any other field, are made by a few, gifted young men and women; their creativity is motivated by challenge. I wanted to supply that challenge, which I had so often missed at Case. Not that the average student should be forgotten. I made the class work hard, but I failed fewer students than many of my colleagues did. Most students prefer a challenging course to an easy one. The average student may not understand everything, but he feels he has learned something if he has to work for it.

I soon found that none of the existing acoustics texts included enough of the newer developments to be satisfactory. So I struggled with inadequate texts, supported by supplementary notes for several years, and then, when I knew how I wanted to present the material, I wrote *Vibration and Sound*. First published in 1936, the book has been through many printings and revisions; it is still being used and being referred to.

The undergraduate acoustics course was fun but my deepest interest was with the graduate students in physics. Compton had come to MIT hoping to strengthen the graduate school. Each of the science departments—and above all, the physics department—was expected to become preeminent in research and graduate teaching. We in the physics department could agree enthusiastically with this aim. We felt it was time the United States began to train its own scientists. We were sure our subject was exciting, and we hoped it would be of increasing practical importance.

Consolidation

Surely every year more physicists would be needed, by governmental and industrial laboratories as well as by other universities. Where was there a better place to train these scientists than in a technically oriented university, such as MIT hoped to be?

As the word of Compton's plan spread, more graduate students came to MIT. When I arrived in 1931 there were about a dozen in physics; within six years their number had grown to sixty-odd. They were a select group, selected in many ways. They had to be highly interested in physics. At that time there was no fortune or fame to be won; achievement in research or teaching would have to be its own reward. In my own case, my yearly salary had started at $3700, and it was not increased much for several years. I am not sure it would have interested me much to have been told that my annual income in 1968 (just before retirement), including book royalties and consulting fees, would be roughly sixteen times that initial year's amount.

Less desirably, the selection of the students also had an economic aspect, for most students had to pay their own way. Each student's parents were thus also untypical to some degree. For many, the drain on family finances for four additional years of education, in preparation for an unremunerative profession, was a serious matter. One vivid memory reminds me of this. One afternoon the father of an undergraduate student I had been working with, came up from New York to ask my advice. He clearly had not had a university background himself and was as clearly uncertain about our world. Omitting his hesitations and apologies, this was what he asked: "My son Richard is finishing his schooling here next spring. Now he tells me he wants to go on to do more studying, to get still another degree. I guess I can afford to pay his way for another three or four years. But what I want to know is, is it worth it for him? He tells me you've been working with him. Is he good enough to deserve the extra schooling? Will it help him?"

It was a struggle to keep from smiling, lest the father think I was laughing at him, for Richard was the brightest undergraduate I had yet met. He was coming to my office each week to learn about quantum mechanics because his scheduled courses conflicted with the regular quantum mechanics course and also because, according to the rules, he was not supposed to be ready to take such an advanced course. He was doing so well in all his courses that the department had taken the unusual step of proposing that he be granted his bachelor's degree at the end of three years instead of four, and we were recommending him for a graduate fellowship at Princeton, I assured the father, as sincerely as I was able, that Richard was, indeed, that good and that he really needed the extra schooling to be able to enter his chosen profession. I must have been persuasive, for the boy went on to Princeton that next year. Twenty-six years later he was awarded a Nobel Prize in physics. I hope the father was still alive then.

So, without conscious planning, I found I had taken on the job of faculty adviser to the physics graduate students. Every spring the applicants for admission to MIT had to be rated. Since there were always many more students applying than we had room for, it was not an easy job to choose among the graduates of many colleges and universities, each institution offering different levels of preparation in physics and maintaining different grade standards. We soon found that letters of recommendation from a student's professors were better gauges of excellence than grade averages, particularly when we knew the professors. I therefore tried to keep in touch with the physics professors in the institutions sending their graduates to us. I started the practice of writing them, telling them how their graduates were doing with us, asking them to compare their next year's applicants to the ones already with us. With our physics colleagues in New England, we could make more personal contacts. Many of them in the smaller colleges felt isolated, having only a few students specializing in physics. Few of them felt it worthwhile to attend the national meetings of the Physical Society, where much of the discussion was about new ideas they had not learned in their student days.

I felt it would be worth the effort to organize some meetings in New England for these older professors to use to talk over their own problems and to learn about the new physics in their own way and at their own rate. I talked it over with Compton, who had been president of the Physical Society a few years earlier, and with Ted Kemble at Harvard, who knew most of the New England physicists. The Physical Society had never had a regional section, but Compton thought the Society's Council might agree to form one, if asked properly. Kemble polled his friends around the region and reported support for the idea. Together we wrote up suggested bylaws and got the Society Council to approve them. Our first regional meeting was in the fall of 1932. Kemble was elected first chairman of the section; I was made secretary. After a few years, I knew almost all of my New England colleagues.

The job of graduate student adviser didn't stop with the admissions process. When the new students arrived in the fall, each had to be consulted to find out his interests and educational background and to reach mutual agreement on courses. Each student's progress had to be checked, to ensure that no one was in over his head. Eventually, thesis topics and supervision had to be arranged, along with examinations.

The students in theoretical physics were my particular concern; they had less chance to know the faculty than did students who worked regularly in a laboratory under some professor's supervision. We arranged a daily tea for faculty and students in a room across from the department library, near most of the theoretical physics faculty offices. In a few years we had established a family spirit. Impromptu parties were held in the evenings in one or another of the students' offices. The long halls of the main building, deserted at night, proved to be ideal for roller-skating. Theater and circus parties were popular. Many of the faculty joined in; quite a few of them were not much older than the students.

Helping to cement the group socially and intellectually were the postdoctoral fellows. The MIT physics department had a series of these National Research Fellows: Bob Bacher, from Ann Arbor, who went to Los Alamos during World War II and later became provost of Cal Tech; Bob Brode, also at Los Alamos and later chairman of the faculty senate at Berkeley; Lloyd Young, later at Rand; Bill Hanson, who would have become the leading microwave expert had he not succumbed to slow-acting beryllium poisoning acquired during his doctoral thesis research; George Kimball, whom I had known at Princeton and with whom I would work during the war. All of the fellows worked with the faculty, but were close to the graduate students in attitudes and interests.

Each member of the faculty gave a graduate course or two. The classes were small, and often became discussion sessions. We tried to present physics not as closed dogma but as an ongoing search, a wriggling canful of unanswered questions, to which the students might help find answers. For a while I taught quantum mechanics, a subject in which it was easy to point out the blank spots on the map of knowledge. I struggled to present the abstract generalizations of Dirac and von Neumann, so different from the simpler formulations I had learned from Kramers and Condon. It was becoming clear that the new formalism was the proper language to use in describing atoms—and perhaps also nuclei. But its vocabulary was new and its grammar was unfamiliar. I found the bright students caught on more quickly than I did; they had fewer preconceptions to unlearn. I verified again that teaching was the quickest way for me to learn thoroughly.

I still felt deficient in the mathematics underlying the new physics; perhaps I could learn it by teaching a course in it. I managed to persuade the administration that it was not inappropriate to have a course in the mathematics of physics given in the physics department rather than the mathematics department.

The first year's course in methods of theoretical physics was a shambles. There was entirely too much to cover, and I was just beginning to understand the material myself. Because I did not yet feel intuitively that the symbol *div* was shorthand for the expansion of a fluid or the outflow of lines of force, I couldn't impart that intuition to the students. It took me as long to devise problems for the course as to prepare the lectures. I wanted the problems to have obvious connections with physics, not to be abstract exercises in equation-solving, but I did not want them to be so realistic as to be insoluble. By the end of the course both the students and I were exhausted, but I had begun to see how it should go.

At times a student's stray question saved me. One Friday I spent the lecture hour showing the class how to calculate the scattering of a wave from an obstacle. I took one of the simpler examples, that of a wave of sound striking a perfectly rigid sphere; a quite similar calculation can be used to find the scattering of electrons from an atom. When the functions representing the wave are fitted at the sphere's surface, the solution comes out as two series, one representing the original wave (the incident wave), unblemished, as though no sphere were present, and the other, the scattered wave, radiating outward from the sphere. At long wavelengths the scattered wave spreads out uniformly in all directions, but as the wavelength is made shorter, directional effects appear. There is a strong beam scattered more or less in the same direction as the incident wave, the part scattered to the side or in the reverse direction being less intense.

I went on to show that the series for the total outward flow of energy of the scattered wave, the amount of energy the sphere diverts from the incident wave, equals the intensity of the incident wave times an area, called the cross section of the sphere for scattering, which is a measure of the fuzzy silhouette cast by the sphere. I showed that this cross section, for sound and for a rigid sphere, becomes smaller and smaller as the wavelength becomes longer and longer; a rigid sphere does not scatter sound waves much longer than its diameter.

At about this point in the lecture I had finished what I had prepared, but ten minutes remained before the hour was up. I quickly decided to show what happens to the series for the cross section when the wavelength becomes quite small. In that case, the wave should behave like light striking a silvered sphere that casts a sharp-edged shadow. I thought I saw how I could sum the series in this case, and I expected the result would come out to be equal to the area of the sharp shadow, the area of the cross section of the sphere itself. But as I was explaining how the series was summed, talking and writing it out on the blackboard, I began to see that it was coming out twice too large; the total scattered energy was going to be the incident intensity times twice the area of the sharp-edged shadow. As I was talking and writing I tried to look back over what I had previously written, but I could see no error. As I wrote the final result, my difficulty became apparent to the class. As the bell rang, I said apologetically, "I seem to have picked up an extra factor of two; I'll show you the correct result on Monday."

I struggled all that weekend, trying to get rid of the extra factor, but it would not disappear. I arrived at the lecture hall on Monday, still trying to decide what to say. Several of the students were already there, and I confessed to them I had not been able to find my mistake. One of them remarked consolingly, "Well, maybe it doesn't have to come out exactly equal to the sphere's cross section."

For some reason this casual remark unlocked my

subconscious, and I suddenly had the answer. The separation of the solution into an unmarred incident wave and a scattered wave—a useful dichotomy for long wavelengths, where there is no sharp shadow—does not describe the short-wave solution, where there is a sharp-edged shadow as well as an outgoing wave, reflected from the sphere's surface. The series called the scattered wave, for short waves, has to do two things: it has to interfere with the unmarred, incident wave to form the shadow; it also has to produce the wave that is reflected outward in all directions. By the time the whole class had assembled, I had the story. The scattered wave, in the shortwave limit, has two parts: one, a beam pointed sharply forward, which interferes with the incident wave to produce the shadow; the other, the reflected wave, which radiates in all directions. In the limit, these two carry equal amounts of energy. Since it is only the reflected wave that gives the sphere's cross section in the limit, the factor two in my calculations was not a mistake; it should have been there.

Obvious? Perhaps. But after that experience I went back to read earlier texts on wave scattering and nowhere found the matter mentioned. Furthermore, a few years later I had a letter from a distinguished physicist who had read the chapter on scattering in my *Vibration and Sound*. He had used the series I had written out for cross section for a whole range of wavelengths and had found that its value seemed to be approaching twice the cross-sectional area of the sphere for very short wavelengths. He asked plaintively whether my formula was wrong by a factor of two or whether he had made an error in his calculations. I could by then reassure him and explain why the factor should be there.

Consolidation

When I came to revise *Vibration and Sound* in 1947, I enlarged the discussion of scattering to include this point. Although wave scattering from spheres had been worked out before 1900, this matter had been overlooked, or else no one had bothered to call attention to it. Yet to me it was a piece in the jigsaw puzzle that made up the understanding of waves that I was trying to give the students.

During the second year of the course, I was helped by two postdoctoral fellows, George Kimball and Bill Hanson. They agreed to take the problem sections, which meant that I could concentrate on the lectures and on writing notes, while the three of us together could work out a better set of problems. The interaction was stimulating. Hanson would criticize the wording of the problems and occasionally find errors in the solutions I had worked out, while Kimball would come up with alternative and easier ways of obtaining the solutions. The graduate students in the class also contributed. They were an articulate group, quick to point out errors and to argue about problems. Several doctoral theses arose from those arguments. The mimeographed notes I wrote were used in the course for a number of years. When Herman Feshbach, who had been one of my students, took over the course in World War II, he put out a revised edition. The two mimeographed sets achieved an underground fame in a number of war projects, so, by the time the two-volume printed version was published in 1953, it already had a market. As I write, it is still selling.

In addition to coordinating the admission of graduate students in physics and, once they were admitted, advising them on their course of study, the faculty adviser was also expected to help find jobs for students after they received their degree. During the Depression this was not easy. Although the output of Ph.D. physicists in the United States was still small, so also was the demand. Physics was not a subject chosen by many students, so the demand for teachers of physics was not great. Aside from the big technological companies, such as Bell Telephone and General Electric, few businesses were willing to spend money on the basic research that physicists were supposed to be good at. In the thirties even Westinghouse had just about disbanded its much-advertised laboratory, which did not recover until Ed Condon took over the directorship in 1937.

Compton and Robert Millikan, the president of Cal Tech, worked to reverse the attitude of American industry. What had to be done was to persuade industry that the physicist is a scientific generalist, able to contribute to almost any technological problem. We had good examples, if the industrialists would listen. Sooner or later the growing knowledge about the way atoms work would have practical applications in chemistry, metallurgy, and communications. Our battery of new electronic measuring instruments promised much greater speed and accuracy in guiding production tools and in controlling the quality of manufactured output. In addition to persuading industrialists they needed physicists, we also had to persuade more high school and college students that a career in physics would be enjoyable as well as economically viable.

As the decade of the thirties proceeded, this twin sales effort gradually had effect. The number of graduate students specializing in physics grew. More important, a larger fraction of the very brightest science students chose physics rather than chemistry or engineering. And a fair number of them went on to make names for themselves in industry. A few years ago, I made a survey of the careers of all those who had obtained Ph.D.'s in physics from MIT and Cal Tech in the thirties. By the sixties about half of these earlier graduates were professors of physics, many being famous for their research and quite a number being heads of their departments. William Shockley and Richard Feynman, for example, had both been awarded the Nobel prize, and Leonard Schiff was head of the physics department at Stanford. The other half were doing well in jobs that did not involve teaching more physicists. About a quarter of them were in academic administrative positions: heads of engineering departments, deans, or presidents. The other three-quarters, three-eighths of the total output, had taken nonacademic jobs doing research in governmental or industrial laboratories or guiding scientific policy in high-level administrative positions. For instance, James B. Fisk became president of Bell Telephone Laboratories; Ralph Johnson became vice-president of Thompson Ramo Wooldridge; Stark Draper designed and built the guidance systems for NASA space vehicles. This second generation of professional physicists in the United States was to do great things.

Summers were times to catch up on research and to relax with the family. During our first spring in Cambridge, we learned of a farm in Casco, Maine, that took in summer boarders. Will Allis, who was working with me on electron scattering, decided to come along, and the Barrows, whom we had met in Munich and who lived near us in Cambridge, also joined us. Wilmer ("Zike") Barrow was in the electrical engineering department, working on what later would be called microwave transmission.

That summer marked the beginning of my enduring love for the northern New England mountains and lakes. Mountains were new to me, who grew up in flat Ohio. But Will Allis was an old hand at them, having learned climbing in the Alps. That first year we tackled only small peaks near the New Hampshire border. But the feeling of victory and of freedom I experienced when, after slogging along muddy trails and scrambling over boulders and steep pitches, I broke out on top, out of breath, with the clouds just above my head and the high mountains nudging up over the widened horizon, was habit-forming. It was like the thrill coming when, after days of brooding and repeated calculations, I finally solved a problem in theoretical research. But here the effort required was less, and the effort was physical.

Several years later we found a boarding place in Holderness, New Hampshire, between Big Squam and Little Squam lakes, where there was room for colleagues and students to stay, to swim and climb with us, and to talk physics on rainy days. The Squam ridge was easy enough for the beginners, with one good cliff, and just beyond were the higher peaks of the Waterville Valley. And beyond that, but not too distant, were Lafayette and Lincoln and Liberty and Flume, with the glorious connecting ridge whose climb made a perfect, energetic climax to the season. We could all go back to Cambridge a few pounds lighter, a few shades darker, and considerably healthier.

Consolidation

Summer was the time when we finished our research papers, although they were also worked on in spare moments throughout the year. It was in those spare hours, squeezed between class preparation and student advising and administrative duties and family, that I began to face up to my weaknesses and strengths as a research physicist.

It was clear to me that I was no Einstein. I could grasp the essence of a new theory quickly and could take effective part in exploring its ramifications. But I wasn't the one to make the initial breakthrough. This realization came slowly enough to cushion disappointment. There were enough other interesting things to do. I could act as a scout, looking over many areas, choosing those that appeared most promising at the time, bringing to bear research techniques that had been developed in other areas. I could call the attention of others to the potentials of the new area, could help skim the cream of research, and could persuade students to explore further. This skill was not the sort of deep-thinking ability that wins prizes and fame, but it was more in line with my urge to explore and my newfound enjoyment in teaching. And, to be honest, the greatest achievements were outside my capabilities. I liked what I was doing; that was the important thing.

My work in acoustics strengthened my evolving self-appraisal. Discussions with Dick Fay convinced me that, with electronics, measurements of sound that had been impossible before could now be made. I also knew that the theoretical techniques developed for quantum mechanics could be used effectively in extending these measurements, predicting other phenomena, and suggesting other experiments. The theory, buttressed by experiment, could be used to design equipment to analyze, reproduce, and control sound. The results might turn out to be useful, a new consideration for me.

I had begun to worry about the usefulness of my work as I began to see the realities of the Depression. My father was out of work; no telephone rate cases were being fought just then. My mother took in boarders. A year or so later my father worked for the WPA, planning for some building that never was built. He aged rapidly and died in 1939.

At MIT I was helping to discover new things while the country was not able to use the ample resources and technology it already had to feed and house its people. What was wrong? What should be done? Should I stop my research? Should I try to influence the actions of the Hoover administration? To do what? I, like most of my colleagues then, knew very little about the machinery of government. Although my parents and grandparents had been active politically, I had taken little part in politics, beyond voting. My first presidential ballot had been cast for La Follette, and my later voting record seemed to continue on the losing side. Many of my colleagues did not even bother to vote. Even if we knew what to advise the country to do, why should anyone in government listen to pure scientists?

Of course, a few colleagues did seem to know what to do. I was asked several times whether I would like to join the Communist party. But I never was a joiner. The party sounded like a political fraternity, and I had avoided fraternities at Case and afterward. Some socialist ideas made sense to me, but I didn't believe that the conspira-

Consolidation

torial approach was the right way to get them adopted, at least in the United States.

So I went on doing what I was trained to do and what I liked to do, exploring and teaching, but with a glance from time to time at the possible usefulness of what I was discovering. It was unlikely that any discovery in acoustics could help relieve the Depression, but one can never tell ahead of time.

The two aspects of acoustics that first attracted my interest were the scattering of sound and the behavior of sound inside rooms. My experience in calculating electron scattering should, I felt, make the sound-scattering computations easy for me. Now that acoustic measurements were so sensitive that we had to worry about the effect of the measuring microphone on the sound wave being measured, these calculations were needed. Some of the results were included in my *Vibration and Sound;* others were reported in papers. It turned out that some of the calculations involved functions that were not well known and whose properties had not yet been reduced to a set of numerical tables.

Consequently I became interested in seeing that tables of wave functions were calculated and tabulated. In 1934 I learned that the WPA had made some money available to give employment to unemployed mathematicians. A project in New York, organized by Arnold Lowan, was using some of these funds to compute tables of functions. I got in touch with Lowan, and, after discussion, he agreed to have his group compute some of the functions I felt were important. With additional WPA funds I hired some other people to compute tables that Lowan's group was not working on.

Calculations of this sort were still very time-consuming. In the eighteenth century, it was said, Baron von Vega employed a platoon of slaves to compute his tables of logarithms. The advent of desk calculators that could add, subtract, multiply, and divide more accurately and rather more quickly than a person could, made the calculations somewhat less tedious. But, even with desk calculators, reams of paper had to be covered with the results of each step in the computation, and each step, for each tabulated value, had to be checked and double-checked for errors. Nowadays a digital computer can rattle off similar calculations in a few minutes, after someone has spent several weeks writing a program for it and more weeks checking the program for errors. I computed some tables, but I was never very good at it, being too impatient and too liable to make errors.

Some of the people I hired to compute tables were graduate students, many of whom needed a part-time job to stay in school. One other, William Sidis, was not a student. He was an infant prodigy who had grown up, the second one I had known. The first one was Norbert Wiener, who still was a prodigy (although not an infant) when I came to know him. He had gone to Tufts because he was considered by Harvard to be too young to enter. He got his B.A. at fifteen, then was allowed into Harvard as a graduate student, and earned his Ph.D. when he was nineteen. By the time I arrived at MIT he was the most famous and by far the most noticeable member of MIT's mathematics department. He was short and round, with a grey blob of beard that lengthened his round face and with eyes whose confiding look was magnified by strong spectacle lenses. At least once a week he would barge into my office, full of all the new things he had discovered or had talked about in his last lecture. He, too, was worried about the practicality of his work and might finish a disquisition on his latest abstruse theorem on Fourier transforms by asking, with a mixture of pride and uncertainty, "That will have a practical application, won't it?"

William Sidis was not so lucky as Norbert—he had no luck at all. In his youth he had been the living demonstration of the educational methods of his father, Boris Sidis, a psychologist. William knew dozens of languages before he was ten; he was a lightning calculator and could lecture on Einstein or William James or the timetables of the German railways. His father would take him to scientific meetings as an exhibit—until he rebelled and disappeared. That was twenty years before I met him. I never did find out how he lived and supported himself. He had travelled a lot but had finally settled down somewhere in South Boston; he never said just where because his mother might hear of it. When he was brought to me by a mutual friend he had on an oversize cap and a stained overcoat that may once have been gray. He looked as though he hadn't been eating much during the previous weeks. His face was drawn, and he squinted when he talked.

When I told him I had a small fund that could pay him to calculate some tables, he got up and started for the door. After we managed to get him back in the chair, he explained that it upset him to do mathematics any more; he could no longer stand mental work. After a lot more talk, however, he agreed to try the calculations, if I would explain exactly each step I wanted him to carry out, so he wouldn't have to use any initiative or judgment of his own. We put his desk calculator in a small, otherwise unused room, so he wouldn't be bothered by other people, and he would silently arrive and leave at odd times. Each Friday he would come to me with a sheaf of completed calculations, take his meager check, and disappear.

After a week or so I began to notice, as I passed his room, that his desk calculator was turning over less and less often. But his weekly batch of calculations was no smaller, and random checking showed they were still accurate. Evidently his ability for mental calculation was returning; he could do as well computing in his head as with the machine. After a month he began to look better and was even willing to stop and talk if I dropped in on him. He would talk about his collection of streetcar transfers, collected from nearly every city in the country. And occasionally he would talk about a history of the Indians of North America that he was writing.

When Sidis completed the first set of tables, I gave him another set to complete. This time I gave him the equation for the function and a few general instructions. After the improvement he had evidenced, I assumed that he would no longer need the meticulous directions given him the first time. It was a grievous error on my part. I didn't hear from him all day. Toward evening he came to my office, his face drawn and his eyes squinting. He explained that he couldn't go on, that I expected him to think and he couldn't. I tried to undo the damage and promised him detailed instructions as before, but all he did was to look down and squint and say, "Well, I dunno." He left that evening and never came back. A year later I heard he was dead.

Consolidation

Another branch of acoustics I explored was room acoustics, which has to do with the way sound builds up and dies out in auditoriums, lecture halls, and ordinary living rooms. It also has analogues in quantum theory. Sound in a room consists of standing waves, excited by the sound generator and absorbed by the material of the walls, floor, and ceiling, and the waves for the bound electrons in an atom are also standing waves. The problem I worked on was to express mathematically the reaction between the standing wave of sound and the material of the wall, so it would be possible to calculate how fast the wave dies out.

The original work on room acoustics had been done by Wallace Sabine at Harvard before World War I, long before electronics. Sabine used a pistol for the sound generator and his ear for the sound detector. He expressed the sound-absorbing quality of a room in terms of a quantity he called the wall's absorption coefficient, but his methods of measurement of the quantity were necessarily primitive. By the thirties, with all the new methods of measurement, it seemed to me that the time had come for parallel advances in theory. Sabine's absorption coefficients were just numbers, not clearly related to the texture and structure of the wall; a more detailed understanding was needed of the way the standing sound waves coupled to the wall.

I talked with Dick Fay and visited the people at Harvard who were continuing Sabine's work. That group was headed by F. V. Hunt, a professor in the department of applied physics. He and I got along from the start. Hunt was voluble, friendly, with a strong competitiveness usually kept well hidden from others and perhaps even from himself. I didn't mind that quality. I wasn't going to compete with him experimentally, and I was reasonably sure I could keep ahead of him on the theoretical side.

We developed the habit of meeting every other week to argue about measurements and concepts. Was it possible to express the coupling of wave to boundary in terms of a wall impedance, analogous to the impedance of elementary loads along a transmission line? If so, how could the impedance be measured? And how did its value relate to the absorption coefficient of Sabine? Every so often, Hunt would surprise me with a measurement he had not let me know about in advance. Most of the time it checked nicely with the theory I was gradually putting together. I came to enjoy the incentives of the muted competition, and Ted was great fun when he relaxed. One of his group, a graduate student named Leo Beranek, was particularly helpful.

By 1937, acoustics at MIT was expanding. I published several papers on sound scattering and one with Draper on pressure waves inside internal-combustion engine cylinders. In the fall of 1939, Richard Bolt arrived; he made an impressive addition to the MIT team. He had started to study in architecture, switched to work with the strong acoustics team at UCLA, and chose for his postdoctoral fellowship to come to work with me. Bolt was slightly built, energetic, articulate, and an imaginative experimentalist; his wife Katherine, was gracious, canny, and enlightened.

Now we had the nucleus of a strong acoustics group at MIT. Fay and Bolt could lead the experimental research; I could supervise the theoretical work. Bolt made an excellent teacher and thesis supervisor. Together we turned out a half-dozen graduates who have become leaders in the field. With the help of Hunt's Harvard group, and notably that of Beranek, we settled the question of acoustic im-

Consolidation

pedance: what it meant, how it could be measured, how it was related to the structure of the wall material, and how, knowing this, predictions could be made about the behavior of sound in rooms and in ducts. At MIT the thesis of a hard-working young man from India, Nautam Bhatt, helped to confirm our duct calculations. By 1940, with six or eight basic papers published, we had skimmed the cream of this segment of science and could turn over the findings and formulas to the engineers to use in controlling sound, which they have been doing since then. It was a pleasant feeling to see the results of one's vagrant curiosity turned so quickly to practical use. My next interest, however, was in a subject with no imaginable potential for application, or so it appeared.

Several years earlier my attention had begun to swing to the newly blossoming field of astrophysics. To understand the metabolism of stars might not have practical applications, but it had tremendous imaginative challenge. The nearest star, the sun, is nearly a hundred million miles away, and the others are infinitely more remote. We can receive on earth but a minute part of all the various kinds of radiations the stars send out. How could we deduce from these frail signals the structure of the stars, the nature of their variety, and, most cryptic of all, the source of the torrents of energy they all pour forth?

Our only hope of understanding these riddles lies in the assumption that the substance of the stars is the same as the substance of the earth, that the stars comprise the same elements, perhaps in different proportions, but obeying the same laws there as here. Naturally these elements are under conditions of temperature and pressure very different from those we can duplicate on earth, but if our knowledge of the way the atoms behave is sufficiently complete, we should be able to extrapolate our earthly measurements to the vastly dissimilar conditions inside the stars. Here again was the need for a theory universal enough in its application for us to be able to deduce, from measurements made on earth, what was happening inside a star.

Certainly classical physics was not adequate for our purposes. It had predicted that a star like the sun would last only a few million years and then would collapse into a cold cinder, but we were fairly sure the sun had been supporting life on earth for a very much longer time. For lack of a valid theory, astronomers had to be content with measuring what they could of the radiations coming from the sun and the other stars, verifying that the elements we know on earth are indeed present there and classifying stars into various species, depending on the relative intensities of the different parts of their radiation spectra.

Henry Norris Russell, whose classes I had attended at Princeton, was one of the first astronomers to argue that the new quantum theory would produce a new level of understanding of these mysteries. Russell himself made great contributions to the first step of the advance, that of finding relationships between the several observed properties of the stars. He and the Dutch astronomer E. Hertzprung demonstrated a regular relationship between the color of a star's light and its total radiant outflow, the star's brightness after being corrected for its distance (which is known for many stars). The bluer the star's light, the more energy it seemed to radiate (there were, of

Consolidation

course, some exceptions, but this was the general rule). This discovery spurred efforts to determine the temperature of the star's skin from the characteristics of its color spectrum. Such a determination turned out to require detailed knowledge of the quantum properties of the various atoms emitting the light, together with an understanding of the statistical behavior of these atoms as they were knocked about in the hot surface of the star.

In 1936, Donald Menzel, one of the promising young members of the staff of the Harvard observatory, came down to MIT to talk to me about atomic collisions. He was studying the behavior of atoms in the sun's photosphere and wanted to know whether some of my calculations could help him. I couldn't be of much assistance because, at the time, I didn't know much about quantum statistics, but the conversations aroused my interest. The following year I began visiting the observatory, made the acquaintance of the small band of dedicated students working with Menzel under Harlow Shapley, the inspiring director of the observatory. Menzel and I arranged a series of informal seminars at which Will Allis and I talked about collision phenomena and the Harvard participants described what they were learning about the conditions in a star. I went back to studying the intricate subject of quantum statistical mechanics, which seeks to describe the behavior of atoms under pressure from their neighbors, as they are jostled about by the random motions that constitute a body's temperature.

Meanwhile I had made the acquaintance of the Scandinavian astrophysicist S. Rosseland, who had come to MIT several times to use Bush's differential analyzer. Rosseland had described to me the problem of comprehending the internal constitution of stars. Somewhere, deep in its interior, the star was generating the floods of energy that somehow had to push through to the surface and escape. This outward torrent heated the matter that opposed its flow, pushing out against the weight of all the matter above it. There had to be a balance, at every point, between this outward push, the pressure of the material above, and the resistance of the matter opposed to the flow. Rosseland and others, notably the British astronomer-philosopher Sir Arthur Eddington, had worked out a complex set of equations relating the energy flow, the density of matter, its temperature, pressure, and resistance to the radiation flow (called opacity) that had to be satisfied at every point if the star was not either to explode or to collapse.

Rosseland was trying to use the differential analyzer to solve these equations, working inward from the surface to determine what the star's internal structure must be. His solutions were not very satisfactory because he didn't know in detail how the matter deep inside the star reacted to the radiation and to the resulting temperatures. The greatest uncertainty lay in a lack of knowledge of the relation of the pressure, temperature, and density of the material (called the equation of state) and also of the material's opacity, its resistance to the radiant flow, at the terrific pressures and temperatures that must exist deep inside. At that time, we knew nothing of the source of the energy, but Eddington had pointed out that we knew its magnitude because we knew the star's intrinsic brightness—what is produced must come out—and that we could begin our attack by assuming that the unknown source is concentrated at the center.

Consolidation

The more I thought about it, the more I was tempted to work on the opacity problem. Such work would utilize my knowledge of atomic collision behavior, and it would force me to learn more about quantum statistics. It would be a long job, involving a lot of calculation. There was no hope of duplicating on earth the conditions existing inside the star, so all the theories of quantum and statistical behavior would have to be extrapolated from earthbound measurements. Many of the laboratory collision experiments duplicated the individual velocities to be expected, but these results all had to be combined to obtain the average resistance to radiation at the high temperatures. It turned out, as I was to learn, that the opacity was very sensitive to the presence of small amounts of the heavier atoms, such as those of iron. It was estimated from spectral analysis of their escaping radiation that most stars consist mainly of hydrogen, with a lesser amount of helium and small amounts of the heavier elements. But the relative amounts of the ingredients were not known, particularly in the interior. If the results of the research were to be useful, therefore, calculations had to be made for a wide variety of mixes, as well as for a number of different pressures and temperatures.

These computations engaged my research time for several years. By the time they were completed, Hans Bethe had published his explanation of the source of stellar energy, the conversion of hydrogen into helium by nuclear fusion. His original suggestion, involving the catalytic action of carbon nuclei, has since been extended to include other possible reactions, appropriate for different stars at different stages of their history. Nonetheless, Bethe's basic theory still stands: at the star's center, under hellish extremes of heat and pressure, the fusion of hydrogen can produce the torrents of energy needed to keep the star going for billions of years. With Bethe's figures on the pressures and temperatures needed to produce the fusion, my tables of opacity could be combined with knowledge of the equation of state to begin a series of solutions of Eddington's equations that would eventually yield a realistic picture of a star's metabolism and, we hoped, even its historical development. In the past ten years, with the help of the largest computers, these hopes are finally being realized, and astrophysics is once again an exciting field. But even in 1939 it was apparent that a solid foundation had been laid for an adequate theory of stellar structure.

Bethe had been an assistant of Sommerfeld's in Munich. I had missed him in 1930, for he had gone off to Copenhagen for a year, but I had heard about him from Sommerfeld. When Hitler came to power, Bethe came to the United States, and I had met him a few times at Physical Society meetings. He once told me how he came to develop his theory of stellar energy. It was early in the development of nuclear science. New measurements of neutron and positron production by nuclear bombardment were being published each month, and theorists were trying out different calculations to see if any one of them would fit. The few tries I had made didn't work out, so I had turned to acoustics and astrophysics. Nearly everyone's trials were shown to be wrong by the next set of measurements. Bethe was one of the few who seemed to keep ahead of the data; Fermi was another.

Consolidation

Bethe had a habit that was unique among American physicists, as far as I know: he frequently checked to see if any prizes were being offered for discoveries in various branches of physics. In 1937, Bethe learned that A. Cressy Morrison, president of the New York Academy of Sciences, was offering a prize for the best paper explaining the source of energy of the sun.

A little thought persuaded Bethe that the nuclear reactions he had been theorizing about must provide the answer, so he set out to devise a sequence of reactions, ones that had already been checked, directly or indirectly, by high-energy measurements, that would combine four protons (hydrogen nuclei) into a helium nucleus. Direct combination of protons, one after the other, seemed unlikely in the light of what was then known. But after several tries, Bethe hit upon the possibility that a carbon nucleus was a catalyst, trapping one proton after another until it holds four of them, which then are ejected as a helium nucleus, leaving again a carbon nucleus and giving off energy at various stages of the process. He found that at about twenty million degrees centigrade, a temperature within the range of possibility at the center of a star, the sequence of reactions seemed capable of producing the energy actually known to be produced. So he wrote the paper, was awarded the prize, and immediately became an important astrophysicist. Once in a while a little money, properly applied, does great things.

In the fall of 1939, the New York Academy held a special conference on the internal constitution of stars, and here all the pieces of the puzzle came together. The grand old men of American astrophysics, Russell and

Shapley, were there. Bethe spoke and so did Menzel and S. Chandrasekhar, the brilliant editor of the *Astrophysical Journal*. I presented my newly completed calculations of opacity. It was agreed by the conferees that although the fit of the puzzle pieces wasn't perfect yet, a few more years of hard work by everyone would bring success.

These optimistic forecasts were rendered nugatory by the approach of World War II. After the war it was found that the calculations needed to verify and extend the theory were much more onerous than had been expected. It has been only in the past ten years, since high-speed computers have become widely available, that significant advances in the knowledge of stellar interiors have again been made.

After the 1939 conference I never returned to research in astrophysics, but I later had a mysterious indication that my opacity calculations were, astonishingly, having practical application. In 1943 a friend called to ask if he could borrow my notes and computations of matter under very high temperatures and pressures. Being security-conscious by that time, he said he could not tell me why he was interested. Being security-conscious then myself, I said I understood. He never returned the folder of computations, leaving the only void in my files of old research notes.

I didn't need to inquire about the reason for his interest, because I had already heard that he had been called to an esoteric research sanctum called Los Alamos. It needed no sleuthing on my part to realize that the group at Los Alamos was thinking seriously of temperatures and pressures close to those caused by nuclear reactions in the sun's center and that they wanted to know what resistance various substances would offer the resulting radiation. I heard, much later, that my notes were of some use to the designers of the atomic bomb. One never knows, ahead of time, which research will have application in what field.

In 1939 I could consider myself a success in my chosen career. I had been made a full professor, and my salary, together with book royalties and a few consulting fees, came to more than twice my 1931 income. With the help of my father-in-law we had been able to buy a house in the nearby suburb of Belmont. In 1939 a daughter, Annabella, was born. I had been elected a fellow of the American Academy of Arts and Sciences, of the Acoustical Society, and of the Physical Society, and I had served for three years as secretary of the Physical Society's New England section. My alma mater, Case, had awarded me an honorary degree. I was well known by my fellow physicists and didn't mind being unnoticed by the rest of the country.

But even before 1939, the pressures of the international situation began affecting my thoughts and, to some extent, my work. Almost everyone who had been to Nazi Germany or who knew any of the refugee scientists from Germany or Italy was aware that war would sooner or later break out and that the United States would almost certainly be involved. I had been a pacifist during the thirties and did not want to have anything to do with a repetition of the bloody waste of World War I, but the more I knew of Hitler's plans and actions, the stronger became my conviction that sooner or later we would have to fight. To some degree I felt relieved that it was not up to me to make decisions about American policy toward Hitler. Physicists, mathematicians, and astronomers were a tiny minority in the country, thought of as impractical when thought of at all, and most of us were quite satisfied with our insulation from national affairs. Still, we occasionally wondered what we would be doing if war came.

I could at first imagine only that in the event of war my scientific work might be carried out under the orders of some captain or major, leaving me little say about the purpose of my work. I viewed this prospect with mixed feelings. I had always disliked the military way of doing things, ever since my youthful contact with it in World War I. Moreover, my nature rebelled at having to work on some aspect of physics without having any control over how the results of my work might be used. But still, at that early stage in America's involvement, I recoiled from having to participate in the grim decisions of a war and was half-thankful that I would probably have to follow orders.

It is almost impossible now, after the militaristic imbecilities of Vietnam, to explain to my younger colleagues how I could pass in the late 1930s from heartfelt pacifism to whole-hearted cooperation in preparing for war. As I watched Czechoslovakia and Austria being taken over and Poland being ravaged, I, like most of my colleagues, gradually became convinced that Nazi Germany was a force we must fight and put down. By the time of the German submarine campaign and the Battle of Britain, the great majority of American scientists were ready to contribute whatever they could. More important, most scientists came to feel, as I did, that scientific work for the war effort should not be entirely controlled by the military and that scientists must have a part in deciding, at the highest levels, what direction their work would take. By the summer of 1940 Bush and Compton, together with James Bryant Conant of Harvard, removed some of our apprehensions on these points by persuading President Roosevelt to approve the formation of the National Defense Research Committee (NDRC), which provided a way for scientists to participate in national decisions related to the development of weapons and military defense.

And so the decade of my progress in research and teaching came to an end. It had been a happy, untroubled time of a sort that has never since returned.

6
Application

One January day in 1941, Dick Bolt and I were perched on the bed in a Washington hotel room, listening to Comdr. E. C. Craig. He had started out in a half-whisper, but as he got into the reasons he had called us down to Washington, his voice gradually took on a penetrating, quarterdeck quality. He had just come back from England, where they were having trouble with German acoustic mines, and his job was to equip our Navy with means to counter these mines, in case we ever had to.

The mines, he said, were part of the German effort to cut off Britain by sea while it was being bombed from the air and to induce it to surrender as the French had just done. The mines were dropped from planes into the shallower parts of the coastwise shipping lanes. As a ship passed over a mine, the rumble of the ship's engines and propellers could detonate it. The British had just begun to learn how to sneak past the German magnetic mines by cancelling the ship's magnetic field by carefully adjusted coils. Now the Nazis had sprung this new variety on them. Since it was impossible in practice to eliminate ship noise, these new mines couldn't be "blinded." They had to be swept. Some device had to be pushed, or towed to the side, over the mine, one that sounded like a ship but cost less than a ship to replace if it were blown up. To design such a device, we first would have to learn just how ships sounded under water. Then we would have to make an

Application

acoustic imitator, louder under water than a ship and smaller than a rowboat.

As the commander began to detail the kinds of measurements he wanted us to make, there came a knock on the hotel-room door. A man of slight build and intellectual features, in a Navy lieutenant's uniform, came hesitantly into the room. "I feel I should tell you," he recited apologetically, "that I could hear you quite clearly in the next room, and I suggest . . . ." He ran down then, but he didn't need to continue. Dick and I watched Commander Craig oscillate between irritation at being chided by a junior reserve officer and embarrassment at being careless about military secrets—all in front of civilians. It was hard for us to keep straight faces. The lieutenant backed out, red-faced, but his visit had dampened the conference. We arranged, in muted tones, to meet next morning at the Bureau of Ships, in the "temporary buildings" on Constitution Avenue.

Only fifteen months before that 1941 session, I had taken part in the conference on stellar interiors; naval problems were furthest from my mind then. Less than twelve months before, I had revisited Case and talked acoustics with Dayton Miller, certainly unaware that I would soon be measuring the underwater sounds of ships. But in that year we had been jerked from the cozy seclusion of pure science and thrust into the hectic world of politics and military needs. Hitler's panzers had invaded Poland in September 1939 and England and France had declared war on Germany. Still, nothing much had happened in Western Europe that fall. Many of us still felt detached. But as the winter wore on, I, like many others, felt a growing uneasiness. Then, in April 1940, Norway and Denmark were overrun, and, by May, Belgium, the Netherlands, and France were invaded. By that time few of us could think of carrying on as usual; we were ready for someone to tell us what to do.

Radar was needed, of all sorts and for many purposes. The British were already using it to detect enemy bombers as they came over from France; it gave warning early enough to tip the balance in the Battle of Britain. The British were beginning to put it in their planes, in addition, to locate ships and surfaced submarines. British scientists had developed the magnetron, which produced sufficient power for the faint echoes from planes and ships miles away to be detected and measured. But the wavelengths of those early sets were too long. Their directionality was poor, and the display devices were hard to read. The whole system had to be improved, and our British friends were glad to have our National Defense Research Committee take over the problem. More powerful magnetrons had to be developed for shorter wavelengths; better display devices and more directional antennas were needed.

Here was a field in which the usual radio engineer's experience was a handicap. At wavelengths of a foot down to an inch or less, things were much different than for the familiar radio waves a hundred to a thousand feet in length. Instead of thinking about currents running along wires, producing fields outside the wires, a person working on radar had to think primarily about the fields themselves. These waves could be run through pipes—in fact, they behaved more like sound waves than like the longer radio waves. The radar project was an opportunity for persons with the flexibility of the research physicists to con-

Application

tribute as much as the more specialized radio engineers, and perhaps more.

It was an opportunity that many members of the physics fraternity grasped in the next months. The first conference with the British team was like a physicists reunion. The Radiation Laboratory at MIT, set up in 1940 by NDRC to carry on radar development, grew rapidly. In an atmosphere of increasing urgency, dozens of the country's best physicists set aside their pure research, left their universities, and came to Cambridge. When they arrived, they had to learn, very quickly, the politics of group projects and the intricacies of military secrecy. There were difficulties: one soldier on guard duty, overzealous in protecting the nation's secrets, shot a staff member who got impatient at having to show his pass so often. But the group members were intelligent people; they learned fast.

I joined the Radiation Lab early in 1941, when it became obvious that some of the ideas used often in acoustics could also be used in connection with microwaves. The scattering of sound waves, their transmission through pipes, and their radiation from horns and reflectors all had their analogues with short radio waves. I helped pass on some of this knowledge, but I soon felt restless. The Lab soon became too big, and it seemed to me that there were some purely acoustical problems to solve, wherein the groups at Harvard and MIT could contribute their talents and expertise.

The first opportunity of this sort came from Fred Dent, an officer in the Army Air Force, who had taken my course in acoustics at MIT. He wanted to know whether we could study the effect on the crew of noise in military aircraft. The study was to be carried out in two parts: first, measuring the noise inside the plane's cabin and devising means to reduce it, and, second, a physiological-psychological study of the effects of the noise on the crew, to see whether it should be reduced. Ted Hunt, Leo Beranek, and I started the project and began recruiting personnel. We went out to Wright Field to fly in a new bomber, the B-17 Flying Fortress. Our flight took place in a snowstorm, at a time when instrument landings were not routine. Later we chartered a commercial airline DC-3 and flew around New England measuring the noise at different speeds and at different places in the cabin. I long cherished the ticket for that flight—Boston to Boston, $560. Later, when we agreed to look into noise inside armored vehicles, we endured another memorable ride, this time in a tank driven at top speed along a country road, scaring the wits out of the automobile drivers we passed.

We found space for the noise-control project at Harvard, with the measurement and control part of the project in the basement of the Applied Physics Laboratory and the psycho-physiological part in the basement of Memorial Hall, where propeller roar seeping out of the Hall's foundations must have mystified evening pedestrians. We recruited two Harvard faculty members to head the two sections, Leo Beranek for the former and S. S. Stevens for the latter. Since NDRC was busy getting itself organized, the operating arm of the National Academy of Sciences, the National Research Council (NRC) was asked to be the middleman, a role it had already learned to play in World War I. The NRC took funds from the Army and supervised

Application

the work, under a contract with Harvard. Hunt and I became part of the NRC supervisory panel, and we went on to look for other opportunities to contribute to the war effort.

It was then that Commander Craig turned up, with his problem of the acoustic mines. He had talked to Compton in Washington, who referred him to me. I called Dick Bolt, who had just left MIT to join the faculty at the University of Illinois, and we went down to our hotel room session. (That was the last Illinois saw of Dick.) The result was a Navy (and later an NDRC) contract with MIT.

Even now I am somewhat appalled at our bravado. Only one of the people we first recruited had measured sound under water, and none of us was aware of the complications of directing naval vessels to run over an array of underwater microphones, so as to measure the sound the vessels produced. In a passing spell of self-doubt we did go down to the Naval Research Laboratory to ask advice, but we found that their experience had been almost exclusively concerned with sonar, which used equipment sharply tuned to ultrasonic frequencies. They knew no more about designing a microphone and amplifier to respond to everything from 10 to 30,000 cycles than we did.

Some of the parts of our task did stagger us a bit. We were assigned a Navy tug, and a section of Massachusetts coastal waters off Nahant was set aside for our first underwater range. We set out to recruit people with the various skills we felt we needed; Dick Fay, my valued collaborator in applied acoustics, was one of the first people we asked. He was a deep-water sailor, and he had a house on Nahant.

We found Robert Lowell, a Navy captain retired because of serious heart trouble, who wanted a piece of the action, whether or not it took a few years off his life. We recruited several electronics experts and a high-fidelity-recording specialist, as well as a number of the brightest and most enthusiastic graduate students we could find. (One of them, J. E. White, became director of the project by the end of the war.)

Recruiting wasn't hard then. Nearly everyone wanted to contribute to the national defense and was pleased to be asked to join. It was a heady time. With sympathy for England, the realization that physicists could make contributions, and the growing certainty that our country would soon be in the war, people put aside jealousies and personal competition. Running a project was immensely simpler with this kind of cooperation.

Within a month our group had designed and built two underwater microphones, using pressure-sensitive Rochelle salt crystals. We had also found several hundred feet of scarce underwater cable and asked the Navy to get more for us on priority. We were ready for the tug to take us out in the harbor to test out the gear. Four of us appeared at the dock one chilly morning, with batteries, amplifiers, meters, and microphones attached to the cables. There we met, for the first time, the lieutenant who had just been assigned as our liaison officer. Bolt and I didn't know whether to be pleased or uneasy when he turned out to be Lt. Richard Parmenter, who had interrupted our hotel-room meeting with Commander Craig. Parmenter must also have had mixed feelings; he was very formal in his manner that day.

We scrambled aboard the tug and, within an hour had

reached a spot near enough the main channel that we could expect our microphones to pick up propeller sounds from passing ships. We turned on the amplifiers, attached one of the cables, checked out the circuits, and threw the microphone overboard. Immediately we could hear clanks and propeller swishes, loud and clear. We all grinned and even the lieutenant smiled. But then the loudspeaker stuttered, gurgled, and died. We quickly hauled the cable up; the microphone casing had somehow leaked. Our smiles were rather strained then, but we had another microphone. We carefully connected it to the amplifier, meticulously tightened the seal screws, and diffidently lowered it over the side—with the same result. Beautiful sounds for about a minute, a gurgle, then silence. The lieutenant wasn't smiling anymore, nor were the rest of us.

Back at the lab, we opened up the casings. Inside each was a solution of Rochelle salt in seawater, instead of a dry crystal. It took us until three o'clock the next morning to find out the mistake we had made in designing the water seal, and it took a week to machine another pair of cases. These worked beautifully. We never had any trouble about water leaks after that one debacle.

When we knew Dick Parmenter better we found that we need not have been so apprehensive about his reactions. He had been in charge of the hydrography program at Cornell and was quite aware of the upsets that can occur in any research program. By the time he was transferred to head a mysterious antisubmarine activity in 1943, the project had come to value his understanding and cooperation.

Our next problem was to find a means to record the

sounds our microphones picked up. Recording overall sound intensity was easy, but we needed also to reproduce accurately every detail of the sound, so we could play it back repeatedly for frequency analyses. A few years later we would have used magnetic tape, but its production was in the future. Any mechanical method had severe frequency-range limitations for distortionless reproduction; the deficiencies had to be meliorated by a system of filters and frequency changers. To push the weak signals from the crystal through long cables, we put a preamplifier out at the microphone, and, since transistors were not then known, a vacuum tube had to be inside the case at the end of the cable. This in turn meant using a five-wire instead of a two-wire cable, so the Navy had to exercise its priority purchasing power again. Calibration was another problem. But this, too, was solved without much delay. We began to feel, as did our colleagues in radar development, that with enough money and with the knowledge gained from the last decade's research, we could build equipment to perform almost any required task.

We tested the system out on the range near Nahant and were satisfied. Next came the more practical problems of getting a ship to run exactly along a preset course, so we knew at every instant where it was with respect to the microphones. This exercise took longer than it took to get the equipment ready and was equally full of mistakes and misapprehensions. We soon found that the Nahant range was too close to shore for the larger ships, so we had to survey in a number of different ranges around the outer harbor. Communication between our ship and the recording team ashore turned out to be a serious problem. After having a destroyer tear down the course before we were ready to record and then vanish in the distance without heeding our pleas for a rerun, we worked out a procedure for putting an ambassador aboard the ship. He could explain the reason for the run, reassure the captain that the course could be run without endangering the ship, and also manage to calm any impatience when we wanted a second or third run. In addition to these diplomatic duties, our envoy could keep in touch with shore by a two-way radio, ensuring that the run was in line with the marker buoys, while we triangulated the ship's position with theodolites and correlated their readings with the sound recordings. James Hopkins, a cheerful and tactful fellow and a former MIT graduate student, was ideal for this assignment.

Our most impressive performance came when the battle cruiser *Quincy* was held over a day to run our range off Gallups Island. The course was long and straight, but it came very close to the shore, where we had set up our telescopes and signal gear; the amplifiers and recording equipment were in a shed above the beach. The first run was at slow speed, but the second was at twenty-four knots. The captain didn't hesitate to follow the course exactly. To us on the beach it looked as though the ship passed us no more than thirty feet away; the bridge seemed nearly overhead. The bow wave nearly drowned everyone on the beach. Off went the *Quincy* to the Pacific. Just ten months later she was sunk in the Battle of Savo Island.

To give the project more flexibility in moving from range to range, the Navy assigned us the *Galaxy*, a ninety-foot motor yacht loaned for the duration by a patriotic citizen. We mutilated the paneling in the aftercabin by fastening up our equipment. Things aboard were crowded during a working day, since naval regulations required a crew of ten, instead of the four that sufficed in peacetime, and the project usually had another dozen aboard to take the measurements. All this added topside weight, which made the *Galaxy* less seaworthy than she had been originally. This was brought home to us when we tried to go via the Cape Cod Canal and Buzzards Bay to the submarine testing range south of New London. We had measured a surfaced submarine in Boston Harbor, but we felt we should measure the sounds from one running submerged. We reached the mouth of Buzzards Bay early one morning and met a sea that set the *Galaxy* to pitching and rolling so much that even the Navy crew was seasick. No one was very sorry when we ignominiously ran back to the quiet waters of Boston Harbor. There the *Galaxy* usefully served both our project and another at Harvard until the end of the war.

After the summer spent measuring ship sounds, it was time to devise and test some underwater noisemakers to see which could mimic the sound of a ship well enough to fool an acoustic mine. This turned into a full interproject competition. The Naval Research Laboratory had prepared several devices, a number of Navy officers had suggested others, and, of course, the British had one in use, an air-driven riveting hammer inside a huge bronze drum. Most of these devices were heavy and hard to handle. Since it is not wise for the ship towing the noisemaker to go itself over the mine, paravanes swing the device out to the side. Paravanes are temperamental, particularly in a rough sea, so it seemed to us that the ideal noisemaker should be simpler and lighter than the existing contestants.

Application

We first thought of generating cavitation noise, which is caused by some rough surface (such as a pitted propeller blade) moving fast through the water, causing the water alternately to pull away and then smack back against the surface. The result, a crackling hiss sounding like frying when produced by the propellers, often is the loudest component of ship noise. We thought of an hourglass shape, open at both ends and towed from one end. Water would be forced through the narrow neck at a speed high enough to pull the water away from the flared aft jet nozzle, causing cavitation. But this device would be bulky and expensive to make. Someone suggested that a pipe, towed sideways might produce cavitation; then someone else commented that two parallel pipes would probably be better, since the water forced between them might be more likely to cavitate.

By this time it was the afternoon before the first day of competitive tests, and we decided we might as well try out this last idea. We asked our mechanic to bolt together two parallel lengths of pipe, rigged so they could be towed crosswise. "Very good," he said; "what size pipe do you want?" We looked at one another. One of us said thoughtfully, "Oh, about an inch and a half in diameter." "O.K." responded the mechanic, when nobody contradicted this, "how long do you want the pipes?"

Answering this took a bit longer, while each of us was moving his hands different distances apart. Finally one of us said cautiously, "Maybe about four feet long," and no one disagreed. "And how far apart do you want the two pieces?" asked the mechanic. We were getting into the spirit of the game by now, so one of us said, with very little hesitation, "Let's bolt them about a half-inch apart." So that was how the MIT entry was designed.

The next day's run was memorable for several reasons. We used the Gallups Island range, with one microphone placed in the center of the channel. Our tug, the *Wapasha*, would start each run a mile or so away, radio us the code number of the gadget to be run next, then tow it at the prescribed speed straight through the course. The tug had some trouble with the Navy's copy of the British "hammer box." The compressed air hose to the pneumatic hammer got tangled up with the towing cable, and, after that got fixed, the whole barrel-shaped object kept careening from side to side as it came down the range. But it produced a satisfactory amount of racket.

Several of the other test devices produced hardly enough noise to be heard above the tug's propellers. Then, on the next run, before the tug got anywhere near the microphone, a raucous buzz began, louder than anything we had heard so far. At first we thought it was some extraneous interference, but the noise got louder as the tug came closer. We checked the code number and found it was our parallel pipes. This was no cavitation sound; it was the noise one gets by blowing on a blade of grass held between one's thumbs, raised to giant proportions. We asked for a rerun, just to make sure. The racket was just as loud. It looked as though we had a noisemaker, but we weren't sure why or how.

A few more entries were run, with only moderate results. And then the tug crew manhandled the last monstrosity over the stern. This was a pair of ship's propellers, fixed at each end of a four-foot axle, with the axle to be towed lengthwise, leaving the propellers free to turn and, presumably, to produce cavitation. As the tug came nearer, we didn't hear anything for a while; then a rhythmic, grating noise warned us that the thing was dragging along the bottom. Someone shouted, "It'll catch on the microphone cable!" But it was too late to do anything about it.

I wish I still had the recording of that day's run. We played it back many times, but it eventually was turned over to the Navy and disappeared into the vast wastebasket of secret records. Partway through was the strangely loud rasp of our parallel pipes. But the climax of the recording came at the end—a crescendo scrape, *scrape,* SCRAPE . . . and silence. Then the recording technician remembered the usual coda, announcing the run number and date for identification: "Aw hell—this is run number 289, November 6, 1941."

A final footnote: when the fatal propellers were hauled back aboard the *Wapasha,* there was our microphone, attached to twenty feet of cable, which had been neatly cut and wound about the propeller axle.

We brought our pipe device back to the lab to see what it was doing to make so much noise. When we saw that both pipes were burnished bright along the lines of closest approach, someone remembered that cylinders towed crosswise generate oscillating turbulence, as witness the wind whistling past a flagpole or the humming of telephone wires. We had just happened to choose the right diameter and length of pipe to reinforce these oscillations, driving the pipes into violent oscillations, beating against each other to produce a monstrous razzberry. They generated the greatest underwater noise for the weight (or for the money) of anything we ever tested.

As far as I know, our Navy seldom used them for minesweeping, although the Canadians adopted a modified version called the Foxer. The Germans had pretty much given up acoustic mines about this time, although we didn't know it then—our Navy never told us. However, as a final twist to this serendipitous tale, those buzzing pipes eventually turned out to serve another, more urgent, need, as will be told later. In addition, of course, we had all the measurements of ship noise; these would also turn out to be useful elsewhere.

By the end of 1941 some of the glamour had left the project for me. True, we had done what we had been asked to do, and had done it quickly. We were proud to be able to show what scientists could do. Most of us liked to build equipment and were enjoying the chance to put to use new knowledge and techniques amassed during the past fifteen years. However, some of us wondered whether the only thing trained scientists were good for in a war was to do the measurements and design work thought up for them by the supply departments of the armed services. Having become acquainted with many of the officers in charge of projects, I entertained a faint doubt as to whether these officers were always asking us to do the right things. Some of us had the temerity to believe we could also assist in deciding what equipment should be built and how it should be used. I felt that our team would be useful on other jobs for the Bureau of Ships, but I personally wanted to get closer to the operational decisions.

My own talents were more theoretical than experimental. Wasn't there a place for one trained to find a pattern in a heap of data, an opportunity, for example, to work out the optimal tactics for new weapons?

Of course, Bush and Compton and Conant, who were

Application

running NDRC, were in touch with the highest military authorities and were undoubtedly contributing to the overall plans. But there must be need for more detailed studies, for mathematicians and theoretical physicists to work at lower command levels, to forge multiple links between new technology and military requirements.

Compton had recruited Jack Tate, professor of physics at the University of Minnesota, editor of the *Physical Review*, and an old friend of mine, to be the top NDRC liaison man with the Navy. We reported to him a few times, and I remember sitting with him after the formal session, trying to explain my feelings of frustration. He understood, but the tight Navy security rules kept him from suggesting anything. The Navy always had been apprehensive about trusting its secrets to civilians, and Adm. Ernest J. King, the new Commander in Chief and Chief of Naval Operations was particularly strict. So Tate had to go slow in suggesting that civilians be brought in on naval operations and plans. Pearl Harbor brought no change in the Navy's position. Then the U-boats began to attack along our coast.

In early March 1942, I was on a ferry from Delaware to Newport News. We passed a tanker that had been the victim of a submarine attack, limping in to port with a ten-foot hole in her hull near the bow. I wondered then who was analyzing the crucial U-boat threat.

7
Invention

Late in March 1942 I was requested to call on Capt. Wilder Baker at the First Naval District Headquarters in the North Station building in Boston. Captain Baker impressed me as soon as I entered his office—steel-gray eyes, gray hair, a look of decisiveness. We talked a little about what his unit was doing; I said a bit about what I was doing—all in generalities, of course, to keep from breaking security. After the captain had sized me up, he began to tell me more details of his new unit.

It was called the Antisubmarine Warfare (ASW) Unit and was part of the Atlantic Fleet, an operating, not a supply, part of the Navy. It had been set up by Admiral King to study and coordinate the defenses against German submarines, which were than having a field day along our East Coast. Baker had spent several months in England, observing the way the British Navy and Coastal Command worked together to protect their shipping from U-boat attacks. He had been impressed by the help provided by civilian analysts in divining enemy tactics and in assessing the adequacy of defenses against them.

Baker had talked with my British friend Pat Blackett, who had helped make radar work for Fighter Command before he became scientific adviser to Admiralty. Blackett had told Baker that the early-warning radar did not become fully effective until some of the civilian laboratory men went to the radar installations to work with the military operators. These "scientists in the field" helped the military find out what the radar could do under wartime conditions; they themselves also learned at first hand the inadequacies of their first designs when used by nontechnical personnel. This experience had persuaded Blackett and other British scientists that new equipment would not be used effectively unless the users were adequately instructed by the designers and unless care was taken to fit the equipment to the capabilities of the users.

Given time enough, the military would, of course, have learned by themselves how the new gear worked and would have been able to tell the engineer-scientists how to improve the man-machine interface. But there hadn't been enough time for this traditional process to succeed. The Battle of Britain had been too urgent to be trusted entirely to the military.

When Baker had returned to the United States he had talked with Jack Tate, the NDRC liaison man with the Navy, about finding some civilian scientific help over here. Remembering my plaints, Jack had suggested me. So at the end of this long introductory explanation—the longest I would ever hear him make—Baker asked me if I would organize a scientific task force to help his unit analyze the U.S. antisubmarine effort.

It didn't take long for me to accept; this seemed to be the opening I had hoped for, and Baker and his staff were men I would be glad to work with. My respect for them grew as our work progressed. These were line officers, accustomed to running ships and planning operations. They were aggressive and, I was to learn, competitive when it came to relations with the Air Force. But, after all, they had chosen a military career, which requires aggression, and they had been promoted for their ability to command. They believed in their mission, but they also knew their place in the traditional hierarchy of American government.

This group of officers differed from today's generation—and from many of their colleagues in the Air Force then—in their attitude toward their civilian superiors and toward the nation in general. They had learned to live with the tradition that they were the servants of the nation, and they often were willing to listen to suggestions from civilians. It was only during and after the Korean war that I began to see evidence, among the younger officers, of the creed that the military are always right, even in matters of political strategy, and of a tendency to view the nation as the servant of the military.

Baker was willing to give me a chance to show what a civilian task force could do. To let nonmilitary persons participate in even minor operational decisions was, of course, heretical to many officers, especially those in the Navy, with their tradition that the commander of the ship or the fleet was absolute master. But Baker had seen enough, in Britain, of the urgency and complexity of antisubmarine warfare to convince him that traditional policies were inadequate here, and he was forthright enough to persuade the old-liners on Admiral King's staff to let him try us out. He never said so explicitly, but it was soon apparent to us that he had put his career on the line; if our group didn't pay off, Captain Baker would never be Admiral Baker.

My first job, as usual, was recruitment. This was getting harder, for more and more of the best scientists were

Invention

being absorbed into defense projects. Near Baltimore, a large group was putting radio into cannon shells to get proximity fuses; in Pasadena, another group was trying to get range and accuracy out of rockets; in Cambridge the Radiation Lab was developing radar. And, in New York, Chicago, and Berkeley another, more secretive group was beginning to look into the potentialities of the power inside the atomic nucleus. None of my friends in these projects broke secrecy rules, but the sorts of people hired by each group, the requests for equipment or advice, and the nature of the small talk when we met made clear to a technical man the general nature of each project, even the nuclear one.

As each of these projects found its task and began to expand, it would raid the others, as well as recruiting from the shrinking reservoir of uncommitted scientists. My ASW project was starting late, so I also had to raid, sometimes knowingly, sometimes unknowingly (if the desired recruit had been hired by another project since I had last heard from him). The protocol was to talk to the prospect privately and informally, to see whether he was interested. If he was, he would speak to his project supervisor, who would then try to persuade him to stay. There was no attempt to ask higher authority to command anyone to go or to stay. It wouldn't have worked anyway. The country needed the best efforts of these people, and people don't give their best if forced to work on a problem that doesn't interest them.

I had a good base from which to operate. MIT and Harvard had turned out a fair fraction of the best physical scientists in the previous decade, and I knew most of them. In addition I knew many of the physicists at Cal Tech,

Princeton, and Berkeley. My job was eased because I did not need many experimentalists; I was looking for people with a more theoretical outlook than was needed by most other projects. I could recruit from the borders of physical science; we needed mathematicians, insurance actuaries, and theoretical geneticists, as well as quantum theorists.

My first cast was among my close friends; it was fairly successful. In two weeks three of us were at work; by the first of May there were seven, and several more had promised to come as soon as they could disentangle themselves from the jobs they had. By September we had seventeen. About half of this first group were mathematicians; most of the others were physicists. Most of these first-comers stayed on throughout the war, becoming heads of sections and field offices as we grew. A few couldn't take the pressure and left soon.

We became welded into a team very quickly. In part this was because we came from similar backgrounds and talked the same language. In part, of course, it was because we were in an alien environment, an innocent bunch of academics within a tight military hierarchy, engaged in a life-or-death task. The officers of the ASW Unit were already overworked trying to organize the air and sea patrols to protect coastal shipping and the U.S. share of the convoy routes to Britain and Murmansk. Some of them couldn't see how we could help the Navy do its own job. And most of them didn't know whether to ask us or order us to do things. Were we the equivalent of officers or of ordinary seamen? We came to realize that this ambiguity had advantages as well as handicaps; we could talk man to man with enlisted men as well as with the officers.

Invention

But the strain caused by mixing military and civilian personnel began almost as soon as we arrived, for, despite our having been asked to help, the Navy's concept of our functioning was quite different from ours. We were shown a room full of reports of all actions by or against enemy submarines, real or imagined. I suppose we were expected to file quietly in, to studiously digest all the reports, and once in a while to emerge to deliver some oracular pronouncement, which would then be implemented by the officers—providing they agreed with us. Our reaction to this unspoken assumption was unanimous, although we hadn't had the prescience to discuss it beforehand. We looked at a few reports and talked to some of the officers who had participated in U-boat sightings and attacks. And we said we wanted to think about the problem before we started to read. It must have seemed like procrastination to the officers, but they weren't sure how far they could order us around.

We went into a one-week huddle to work out a theory of the process of antisubmarine warfare, a simplified but quantitative pattern of the action. Our reasoning went something like this. If the submarine is dangerous because it is hard to find, then the process of finding the submarine is an important part of the counteraction. If it is submerged, then underwater sonar may be used to find it. (At the time this could be done only by specially equipped surface vessels.) But German U-boat tactics of that period called for the submarines being on the surface more than half of the time, to charge their batteries, to travel quickly, or to communicate by coded shortwave radio with their base in Germany. The U-boats would attack submerged, but much of the rest of the time they would be up where they could see and breathe.

Therefore aircraft could be used to find them. Airplanes were in fact being used, with visual observers during the day or with air-search radar during either night or day. How far away could a surfaced submarine be seen? And were they always seen? What percentage of the time were they missed? If we could answer some of these questions quantitatively, we could compute how many square miles of ocean a plane could search in an hour and therefore how many planes would be needed to cover a given part of the ocean with a fifty, or eighty, percent chance of spotting any surface U-boat. And we could work out the most efficient search pattern, the paths the search planes should fly over the ocean to find a submarine as quickly as possible. Similar calculations could be made for the sonar-equipped destroyers protecting convoys and searching for submerged U-boats.

But before we could use our search theory, we had to decide how the quantities we had worked out could be evaluated. Some of them, such as the range of vision of the search radar, had been evaluated by the scientists and technicians who had designed the equipment and the companies that had manufactured it. But were their "test stand" ranges the ones to be used when a radar set was being operated by a tired crewman, being flown for eight hours over a rough ocean in a noisy plane that had none of our present navigational equipment? (The crewman's feelings about the equipment were relevant too; rumor had radar sets producing "dangerous radiations.") We needed to determine real operational ranges that could be used in the search equations to foretell correctly how effective a search plane could be under wartime, rather than laboratory, conditions. These values would have to be obtained from the operational reports themselves, from the reported ranges and relative directions of first sightings of actual U-boats.

Thus the results of our intensive analysis were a set of definitions of important quantities and equations relating these quantities so as to predict search efficiencies and patterns, as well as to specify a procedure to evaluate the quantities from the answers we hoped to find in the operational reports. Once we had our search theory, we turned to the piles of reports of sightings.

We immediately ran head on into an obstacle that was to hinder us in all our work. The reports failed to answer most of our questions. Whoever had made up the report questionnaire that was filled out after every submarine contact had thought of a lot of things it might be interesting to know, but the heterogeneous information recorded showed that he had given little thought to the basic structure of the operation of search. There was so little emphasis on the quantitative aspects that only part of the data we needed was reported. And the part that was there was suspiciously shifty. Estimated ranges for first sightings by radar seemed to vary capriciously, and most were less than half the distances expected by the designers at the Radiation Lab. In many cases it looked as though the questionee, when he didn't remember the details, guessed rather than saying he didn't recall. If this was all we had to work with, it seemed as if we might as well give up trying to apply any theory of search.

Our reaction to this disheartening discovery was just as unanimous, and just as visceral, as our first reaction. We believed our theory, we didn't believe the reports, and we wanted more data. And this led us to our third instinctive reaction, which uncovered one basic difference between the scientific research attitude and the hierarchical attitude of a military organization. We wanted to get as close as possible to the operation we were studying, not to be given data at second or third hand. We wanted technical data to be collected by technical men.

When we told Baker we wanted to go to the antisubmarine bases to see for ourselves, he countered by suggesting that participants in U-boat contacts be brought to our office to be interviewed. A few experts in these kinds of operations could answer all our questions, he was sure. Later we found logic and examples to substantiate our distrust of this procedure; at the time only instinct led us to veto the suggestion. We wanted to go ourselves. Baker finally agreed and went through the negotiations to get our men accepted at the various bases.

Knowing we were setting a precedent, I carefully picked our first field representatives for their tactfulness and presence. They didn't let us down. By June Philip McCarthy was at Eastern Sea Frontier Headquarters in New York and Arthur Kip was at Gulf Sea Frontier Headquarters in Miami; by July Robert Rinehart was at the Caribbean Headquarters in San Juan (and later on Trinidad), and Maurice Bell and John Pellam, a graduate student, were at Argentia, Newfoundland, learning about Atlantic convoy problems. The initial experiences of all these men were surprisingly identical. In each case, they were able to persuade those who carried out the search patrols that accurate reporting was important. As we had suspected, little care had previously been taken. As one

Invention

pilot said, after he had listened to Kip's talk on the work of our group, "Hell, I didn't think anyone ever read those damned reports."

It wasn't long before our men went out on occasional flights and began to see for themselves the details that had never got into the reports. And almost immediately our needs began to be satisfied. We could fill out our theory of search, putting in numbers instead of symbols and modifying details to make the equations fit reality better. Soon we were able to report back to Radiation Lab some of the reasons why the radar was performing so poorly and to suggest some ways in which the defects could be remedied. Very shortly we could present to the ASW Unit a set of search plans that, when implemented, noticeably increased the number of submarine sightings per week. Data coming back from the bases began to be accurate enough for us to use it to spot changes in U-boat tactics.

Search, of course, was just half the problem; the submarine had to be attacked after it was found. At that time, the only weapon the plane had was a slight modification of the depth charge, originally developed to be dropped astern of a destroyer, to sink slowly and then explode at a predetermined depth, presumably close to the submarine. The weapon was quite inefficient when used by planes. If dropped from higher than a few hundred feet it often broke up on hitting the water. In many cases, when the fliers could see that the charge had exploded near the sub, there was no evidence that it had harmed the craft. It was discouraging, to say the least, for a plane crew to make an attack dangerously close to the surface, within the range of small-arms fire from the U-boat, to find that their only weapon had no effect.

Our field men, when they dug into the details, learned that the depth setting for the air-dropped charges was usually seventy-five feet, which had been found best for destroyer use. In the destroyer depth-charge attack, however, the U-boat was submerged. When a shallow setting was used, the charge exploded too close to the ship. But the airplane was well away before even the shallowest setting exploded the charge. More important, in an airplane attack, the U-boat was usually still surfaced. An explosion seventy five feet below the surface might shake up a surfaced sub, but probably would not rupture it. Shockley, who had come to the group in May on leave from Bell Labs, suggested that all air-carried depth charges be set to explode at a thirty-foot depth, the shallowest possible.

Within two months it was apparent that this change had increased the number of attacks resulting in sub sinkings by a factor of about five. The change was a simple one; it would have been tried by someone eventually. It became obvious as soon as accurate attack data were put alongside technical details regarding depth charges; our group just happened to be the first to want to do it.

These quick successes gave us the beginnings of a reputation, so that Baker could take us with him when his unit moved in June to naval headquarters in Washington. Even so, it took some selling on his part. It must have been traumatic for an old-line naval officer to see, as part of the staff of Admiral King, the Commander in Chief of the U.S. Fleet, an ASW Operations Research Group (ASWORG) made up of civilian scientists who were neither officers nor civil servants and who were not even paid by the Navy.

Admiral King did insist that NDRC, who did pay us

through a contract with Columbia University, order us to work solely for the Navy and to disclose no information even to NDRC except when specifically authorized by the Navy. After being reassured on this point, King left us alone, for he was primarily interested in the Pacific war. He had placed the direction of ASW in his own staff because he wanted to keep the Air Force from taking too big a part in it and also because the operation cut across naval command lines.

I never saw Admiral King until just after Hiroshima, but we soon became acquainted with most of his staff. We moved into back rooms on the third floor of the Navy Building on Constitution Avenue in August 1942, were assigned knowledgeable petty officers to handle office routine, and began to learn what it was like to be imbedded in a military staff. We soon became adept at internal politics. By the end of the war we were veterans at it, since most officers stayed in Washington only a year whereas our central group stayed on. We even found a way to get Admiral King an air conditioner for his office after his staff had tried and failed.

Our recruiting efforts were paying off; there were fifteen in the group by summer and thirty by the end of 1942. Shockley was director of research and had a roving commission. George Kimball, who had left the Columbia part of the slowly developing A-bomb project to come with us in July, became head of the Washington research section, and Abe Olshen, who came to us from the Oregon State Insurance Commission, became manager of the Washington office of the group. Olshen's skill at personal relations soon made him our expert on Navy politics and protocol. Our status was indirectly raised when Captain Baker left to take part in the battles in the Coral Sea and become admiral; his unit was enlarged into a curious entity called the Tenth Fleet, which had no ships of its own but which operated all ASW forces. It was headed by Adm. F. S. "Frog" Low. The nickname referred to his voice, not his manners. He was a wise, hard-headed protégé of King, who kept us in line while retaining our friendship.

We shortly took over the record-keeping for the whole U.S. antisubmarine war. A complete IBM data-processing system was installed; we hired a computer expert and several insurance actuaries to work up the programs to analyze and tabulate the data that were now pouring in. Every morning a special list of all Allied shipping losses, U-boat sightings, and attacks that occurred anywhere in the Atlantic during the previous day was printed out for Admiral King's staff meeting—which we never attended. This whole data-processing activity was guarded with a care bordering on paranoia, partly explainable because this was the first time highly secret information had been handled by a battery of ill-understood machines—who didn't salute and couldn't be shot. Only those of our group who were needed to run the equipment and program the analyses were allowed in the data-processing room; the rest of us had to send in requests for data and analyses. Even the IBM maintenance men were not allowed access; the Navy trained its own maintenance crew.

All these procedures emphasized again the contrast between the military passion for secrecy and the scientific need to know. In preparing the punch cards for each incident, our staff would read and code the action reports.

Invention

But certain crucial data, such as the submarine's location and the degree of damage it had sustained in an attack, were issued to us from a still-more-secret room, which none of our group entered. At first it wasn't clear just what lay behind those closed doors, but a few episodes went a long way toward clarifying the matter, as well as highlighting the ineffectiveness of the secrecy when used against us.

For example, Jay Steinhardt, a new recruit, took on the job of seeing how accurately our radio-direction-finding (RDF) net located U-boats. Each day each enemy submarine talked to its headquarters in Germany—in a burst of high-speed code, of course. A set of RDF stations along our coast recorded these bursts and triangulated the position of the U-boat. Planes could then be sent out to attack it. In some of these cases the sub was found and attacked, so data were available on the actual locations of the U-boats, to compare with the RDF estimates of their positions. Steinhardt found that the compared differences were unbelievably small. The estimates of U-boat positions, given to us from the secret room as RDF estimates, checked the actual positions of U-boats found and attacked with an accuracy ten times better than the RDF equipment was supposed to possess. It sometimes is possible to make a machine perform better than expected. In this case, however, accuracy depended on the wavelength of the radio wave, and waves aren't that obliging; something else had to be enhancing the RDF accuracy.

In one of my daily sessions with Admiral Low I reported Steinhardt's finding, simply saying with a straight face that the reports of U-boat locations given us by the secret room had much greater accuracy than the RDF equipment could possibly produce and that we were going to investigate the anomaly. Admiral Low, also with a straight face, said that was interesting. But the next day he called Steinhardt and me in and disclosed what by that time we had guessed but never mentioned, that our side had broken the German code and that the locations given to us as RDF readings were in fact the positions reported by the submarine skipper himself to his commander in Germany. The whole episode convinced me that Admiral Low really would have let us in on many more secrets, if he had been allowed, but he had to have specific arguments in each case before he could persuade Admiral King to relax.

This verification of our surmise led us to put more faith in the submarine-damage estimates also emanating from the secret room. Since this output represented the U-boat commander's own estimate of damage sustained rather than that of the pilot of the attacking plane, we could use it with confidence to compare the efficiencies of various attack tactics, being sure that the reports of results were not overoptimistic. Before we were sure of the source of the damage estimates, we had hesitated to use them for such comparisons. A plane five hundred feet above the ocean is a poor observation post from which to determine whether a U-boat sinks or only submerges under power. For example, we could be sure the nonchalant report from one Air Force pilot, "Saw sub, sank same," was much exaggerated; that same sub reported the next day that it had just sunk a tanker. Even a destroyer captain may be expected to be overoptimistic about oil slicks or the presence of debris, as he tries to avoid possible enemy torpedoes in stormy weather. However, if damage estimates were coming from reports—or an absence of reports—from the submarine itself, comparison with attack tactics and weapons could be made reliably. It always is dangerous to let the producer judge his own products. In particular, to believe the Air Force estimates of the damage it has caused by its strategic bombing is to risk losing the war.

Admiral Low ordered Steinhardt and me not to tell the other members of the group the source of the secret-room reports. We were amused and irritated. Newspaper publicity about code-breaking would, of course, be damaging; the Germans would change the code. But to keep the matter secret from a group analyzing U-boat behavior meant stultifying the analyses. Technical matters have so many cross-connections that a falsification of one part stands out like a defect in a moiré pattern.

Some of the group were irritated by what they called the Navy's refusal to make us members of the family. They felt that withholding facts from us restricted us to the smaller, tactical problems and made it impossible for us to help in the bigger, strategic decisions. No doubt it did, but we were into the Navy more deeply than anyone had thought possible for civilians, and I felt we could penetrate more deeply still when we learned enough to be able to contribute to the bigger problems. The Navy's attitude bothered Shockley enough, though, so that he eventually transferred to a group of technical advisers to the Air Force staff, organized by Ed Bowles of the electrical engineering department at MIT. As far as I could see, the Air Force had a different technique; they let their group in on all, or nearly all, the secrets, and then paid almost no attention to the group's suggestions on the big questions.

In the end this technique may have been more frustrating than the Navy's.

Our men in the field had many fewer facts withheld from them. Once they became members of the family at the operating base, they were in on any problem they had time to work on. By the fall of 1942, we had developed a pattern of field assignments that worked well throughout the war. No group member was assigned to a field command until an invitation had come from the field commander. We then expressly stipulated that the assigned group member could stay only six months, and that, while there, he would report data to us in Washington only when the commander approved. This procedure went a long way toward allaying the commander's fear that our man was coming as a spy from Washington to check up on the field command.

The sequence of events in "opening up" a new base soon became almost routine. Some junior officer who had worked with us in Washington would be transferred to an operating base and would begin to talk about what we were doing. Eventually an inquiry would come to Admiral King's staff, asking whether one of our men could be assigned to the base. If it appeared that the assignment would be productive, we would send a member out, after filling him full of our latest work and sending him to the NDRC labs developing equipment likely to be used by the base. Within a month of his arrival, his familiarity with new equipment, his knowledge of happenings at other bases, and his ability to analyze local data made him popular and overworked.

Invention

We limited the length of our men's field assignments to six months for a number of solid reasons. We wanted the men to bring back first-hand information about what was happening in the field. Also, new ideas were being generated by our central group, and new equipment was being developed by the laboratories, so that a field man needed to be brought up to date after six months, if he was to remain useful in the field. After six months, of course, the commanding officer at the field base was sure our man was too valuable to lose; if we had not had the six-month rule, our man would have been stuck out there until the end of the war. When information from the base was still important to us, we would send out a replacement, and within a month the new man would be just as irreplaceable as the first one.

But this is getting ahead of my story. About the time the group moved to Washington, Shockley and I decided we ought to visit England. Baker had first got his idea for the group from Blackett at the British Admiralty, and it was about time for us to find out what the British had been doing. I had known Blackett at Cambridge a decade before, and I was curious to know how he had made the transition from physics to war.

An additional reason for making the visit just then was that an Air Force squadron was being sent to England to assist the British Coastal Command patrolling the Bay of Biscay to catch U-boats coming in and out of their bases on the French coast. Attacks on the bases themselves didn't seem to accomplish much; the docks were very heavily protected. But offshore patrols, particularly night patrols using radar, would often catch subs running on the surface to get out to sea as fast as possible. It was a tough assignment, eight hours of night flight close to the deck, watching the temperamental radar screen. Even the day patrols were not much better. The weather was often foggy, when it was clear, there was danger of attack by German planes. The British were short of planes and crews and asked our Navy for help.

The request came at a time when the Navy's tug-of-war with the Air Force was at its liveliest. Just after Pearl Harbor, the Navy had too few large planes for it both to patrol the Pacific and to protect shipping in the Atlantic from U-boats. The Air Force was willing, indeed anxious, to gain a foothold in the ASW effort. Land-based planes could be used effectively for the coastal patrols and could even cover much of the convoy routes, once bases were built in Newfoundland, Labrador, and Iceland. The heavy bombers, especially the B-32 Liberators, had the necessary range and could carry radar and depth charges. The pilots were not trained for an ASW job, especially for sea search and for navigation over water, but perhaps this drawback could be overcome. The U.S. Air Force also saw the Royal Air Force Coastal Command as an example to emulate. In Britain the Coastal Command's chief task was the aerial part of the antisubmarine war, which it carried out through a chain of command quite separate from the Navy's chain.

This arrangement was not to the liking of our Navy. Admiral King and other officers were convinced that naval air—even ASW air—was just a part of a cooperative operation involving ships and planes and that separating off the air arm would interpose so many command links that instant, understanding cooperation would be difficult if not impossible. Having seen later the difficulties between Army ground troops and tactical air, controlled by the Air Force, I am inclined to agree with the Navy. However, at the time there weren't enough Navy air squadrons to take care of all ASW needs, so the Air Force had to be brought in. Six months later the Navy finally had enough planes, so the Air Force was summarily squeezed out of the antisubmarine war. The session between Admiral King and Gen. H. H. Arnold, when this occurred, must have been worth witnessing.

At any rate, Shockley and I in November 1942 flew to London by a rather roundabout path. We went to Lisbon in a Pan American flying boat, with lengthy stops at Bermuda and again at the Azores to refuel and to wait for the sea to be calm enough so we could take off again. We finally landed in the bay at Lisbon, in the evening, after circling for a half hour waiting for the fishing boats to get out of the way. After a day's wait in Lisbon, we were flown at night to London in a blacked-out British transport plane, to avoid being spotted by German fighters from France.

Our visit came after the London blitz and before the V-1 missiles began to come over, so things were fairly quiet, although everything was blacked out and there were occasional air-raid alerts. It took us a while to learn to navigate the sidewalks at night, particularly if it was foggy, with the traffic roaring past almost unseen, a few feet away. Headlights were of less than firefly intensity, and traffic lights were as dim or nonexistent. Luckily we were transported to and from appointments in official cars, usually driven by cheerful, iron-nerved girls in uniform.

Our visit was efficiently managed by Bennett Archambault, the NDRC liaison officer in London, a veteran of the blitz who seemed to know everyone. We spent many days with Blackett, who was even leaner and quieter than when I had known him in 1931. He introduced us to various naval personages and told us how he operated and what he had learned of U-boat habits. Occasionally we would talk about politics and strategy. He already was pessimistic about British Bomber Command's policy of massive night bombings of German cities; he thought the damage would not incapacitate the Germans, but would spur them on to greater efforts, as had been the case in Britain.

We also visited the various groups of scientists recruited to assist the military commands, both British and American. These groups had come to be called operational research groups in England; in the United States we dropped the *-al* in favor of the term "operations research" (research on operations), or O/R for short.

We visited the Admiralty group and the one working for Coastal Command, which was responsible for the air antisubmarine patrol of the Bay of Biscay. We also visited the O/R team attached to the U.S. bomber force in Britain, under the command of Gen. Carl Spaatz. It gave me a chance to see how our Air Force had managed its O/R recruitment. I never found out whether the Air Force simply wanted to emulate our Navy or whether they had independently heard of the British use of scientists at the operational level. At any rate, they had asked W. Barton Leach of the Harvard Law School to recruit some O/R teams. Leach, a very energetic man but one with few acquaintances among the scientific fraternity, had consulted with me several times during the previous summer.

We had agreed on some things and had disagreed on others, particularly in regard to organization. For his teams he had acquiesced in relinquishing all control over his men when they were assigned to a field command. Thus the O/R group working for the Air Staff in Washington had no official contact with or control over the field groups. In contrast, I had by that time persuaded the Navy to return each of our field men to Washington at the end of six months, whereas the Air Force men stayed in place for the duration.

The disagreements between Leach and me seemed to me to be a sign of more basic differences in viewpoint. I looked on the job of my group as being one of technical consulting. We were to look the facts in the face, to measure as many of them as *we* felt were important, and then to report to the commander what we had found—whether or not it ran counter to his predilections—so he could decide what to do about it. Leach's view of the work seemed to be less definite. He felt that an O/R team was simply an addition to a commander's staff, to do whatever the commander wished. The difference may seem minor, but to me it was basic. I had insisted on the right of our group to spend a sizable fraction of our time on self-initiated studies, whereas Leach saw no difficulty in having the military authority approve all studies in advance.

The O/R team assigned to the U.S. bomber force in the U.K. was the first Leach-recruited team I had the opportunity to visit. The head of the group was a lawyer, but among the other members were some I had known before, in particular, my old friend and teacher, Bob Robertson. Bob already was troubled about the plans then taking form for bombing German cities. He felt that strategic bombing was turning into an emotional rather than a rational way of waging war. The Germans had bombed Britain; we would exact an eye for an eye. Bob could already show that this kind of bombing would inevitably destroy more homes than military targets, and he felt strongly that such destruction would strengthen, rather than weaken, the German will to resist, just as it had in England. I agreed with him then, just as I agree with him now. High-altitude bombing is a bully's way of fighting; it is cruel, vindictive, and immensely inefficient, like burning down a lived-in house (with all its inhabitants) to kill an intruding rattlesnake.

But Bob wasn't making much headway with the leader of his group, let alone the staff of the U.S. bomber force. The lawyer heading the group had monopolized the group's contacts with General Spaatz, so Bob's pent-up doubts had to be delivered to me and to his group colleagues, none of whom could do anything effective about them. Then and later, I felt that we at the working level often failed to reach those up in the higher levels of command, to tell them what really was going on. The words they seemed to be batting to and fro up there didn't seem to have much relation to actual events, and therefore their decisions sometimes brought disastrous results. I could only hope that our activity, the work of all the O/R teams, could link high-level decisions with operational events, the words with the realities. Some sort of feedback is needed to prevent a complex machine or organization from running amok.

Working with the U.S. Air Force ASW squadron in Cornwall, Shockley gathered some information that sup-

Invention

ported our instinctive reactions of the previous spring. The squadron had just begun to fly its lonely, eight-hour patrols down toward Spain and back, looking for surfaced U-boats. For a while, the submarines had been using the night to hurry, surfaced, through the danger zone, out to the comparatively safe mid-Atlantic. But then Coastal Command had started flying at night with planes equipped with radar and searchlights; they caught a number of subs, sank some, and scared the rest. As a result, night contacts dropped off and the Coastal Command O/R team believed daytime flying would pay off again. The U.S. squadron was requested to close the daytime gap.

A week after Shockley arrived one crew returned from its patrol excited and frustrated. They had spotted a U-boat on the surface and had come down on the deck to attack. The bombadier had pushed the button to release the depth charges and nothing happened. Cursing, the pilot swung around for another try, dove down again, and again no depth charges dropped. By the third try the U-boat had submerged, and the crew came back to base seething with frustration. They soon found the trouble. In the damp Cornwall air the bomb shackles had rusted tight. Needless to say, that crew greased its bomb shackles diligently before they took off two days later, with fire in their eyes. They found no submarine, the weather fogged up before they returned, they missed the field and ran into a Cornwall cliff. That was the end of that crew.

When we returned to the States we checked Shockley's information and impressions from Cornwall with the records of other ASW units and found that, on the average, the crew of an ASW plane had just one chance at a German submarine during the crew's "active life," before its members were dispersed for other duties or were killed or wounded. An ASW crew therefore had no chance to learn by doing. Fighter pilots and ground troops usually fought several battles during their active life; if they were lucky during the first few "lessons," they learned how to fight by fighting. But if ASW fliers didn't do things right the first time, they usually had no other chance; there was no such thing as an experienced antisubmarine air crew. Therefore the *only* way to determine the best antisubmarine tactics was for a group such as ours to study *all* the reports of battles with U-boats, made by *all* the crews, and then to determine, by comparison, what to do and what not to do. This realization strengthened the conviction of our group that we were needed and that we had to see for ourselves what was going on in the field. And it also deepened our sense of responsibility.

I had planned to come back to Washington before Christmas; Shockley was to stay on another month. All available planes were full of high-ranking officers returning for Christmas, so I was booked on a return trip of the *Queen Elizabeth.* I wasn't told where the ship would sail from or when, but only to be at a specific platform in one of the London railroad stations at a specific time. I arrived there (at night, of course), to find a shadowy crowd of military personnel and civilians milling around, wondering if this was the right place and hesitating to ask. Finally we were herded into a train and started out, whither we knew not. The next morning we found ourselves at dockside somewhere along the Clyde. We went aboard, were assigned our bunks, and then waited again for night.

Invention

Others weren't as lucky; to hear the stories, England must have been dotted with strays, plaintively asking where to board the *Queen*.

My bunk was the bottom of a three-high tier that had replaced the single bed in a once-exquisite first-class cabin; the bunks' two-by-four uprights had been brutally nailed to the costly panelling. Once under way we went at top speed, much faster than that any convoy escort could attain and too fast for U-boats to chase and attack us—we hoped. All through the war the two *Queens* crossed unescorted and were never attacked; later we heard that many U-boats saw but could not catch them. Eastbound, they were packed with American troops; westbound, they carried prisoners of war and a few returning military personnel and civilians.

We ran into rough weather, and even the *Queen* rolled and pitched. As we passed a convoy one day and watched the escort destroyers climb up one wave and dive into the next, I wondered how they could guard against subs in the midst of such gyrations.

I got back just three days after Christmas and started again my weekly commuting between Boston and Washington. By the time I returned, the brunt of the U-boat war was being borne in the Caribbean, where entirely too many oil tankers from Venezuela were being sunk. Our group representatives in the Miami command headquarters were working overtime to keep on top of the radar situation. The new Navy ASW Liberators were equipped with the new 10 cm radar, developed by the Radiation Lab at MIT. These sets had a longer range than earlier radar, allowing the plane's crew to detect a sub before the sub's crew could see or hear the plane. The Liberators were also equipped with plan-position indicator scopes (PPI), which displayed the echoes as on a map, with the location of islands and shipping shown. The PPI made it possible to plan the attack approach—over an island, say, or from out of the sun—so as to surprise the U-boat before it could dive.

It was obvious that the appropriate countermeasure for the Germans would be to equip their submarines with radar detectors, to give warning of a plane in time to dive. Our group, with its contacts at the Radiation Lab, knew that such detectors would not be easy to install or maintain. But the Navy fliers were continually convincing themselves that the Germans did have radar receivers. Our fliers were therefore turning off their radar sets, lest they warn the U-boats before they could catch them. Any report of a disappearing contact, a presumed U-boat echo that disappeared before the plane reached the indicated site, would start another round of turned-off sets.

Since visual search had less than half the range of radar search, shutting off the sets cut our search capacity in half. Until the German submarines were equipped with efficient radar detectors, any limitation of our radar was harmful. Thus one of the tasks of our overworked men in Miami was to keep a close scrutiny of the reports of all U-boat sightings, to catch any signs of a sudden increase in cases of early submergence. A few scattered disappearing contacts would not count; only if more than half the contacts disappeared, it was decided, should our radar be turned off—until a counter-countermeasure was developed. Several times our men managed to persuade fliers to turn their sets back on, by showing that if the Germans had a radar detector, it wasn't working well enough to matter.

Invention

Then came the case of the disappearing blip. One of our more dependable air crews, when approaching the Mona Passage between Puerto Rico and Hispaniola, saw an echo on their scope, indicating a submarine at about forty miles distance. The echo disappeared when the plane was within about ten miles of its target, and no surfaced vessel was to be found. When this happened a second time, in the same region, we had to act. We had a Radiation Lab expert go down to Miami and fly with the crew. Once again, as the plane approached the Mona Passage, a blip appeared at forty miles and disappeared at ten miles from the target. The expert asked the pilot to reverse his course—and the blip reappeared as the plane reached the ten-mile distance again. The mystery was solved; the phenomenon was a "second-time-round" echo.

A radar set of the type then in use sends out a whole succession of pulses. It first sends, then listens, then sends again, and so on, at a frequency called its repetition rate. The rate for that set was about 250 times a second. Radar pulses travel at about 50,000 miles a second, so a pulse could go out and be reflected from a ship 50 miles away in $1/500$ of a second, which would be recorded well before the receiver is shut off in order that the next pulse be sent out, $1/250$ of a second later. If the pulse were reflected from a target 150 miles away, it would return in $3/500$ of a second, which would come in $1/500$ of a second after the *next* pulse had been sent out (*if* the time between pulses is exactly $2/500$ of a second). The pulse would thus appear on the scope as though it came from only 50 miles away; it would be a "second-time-round" echo.

Of course, the echo from a ship 150 miles away was usually too weak to be recorded at all, so these confusing double images were infrequent. But if the object 150 miles away is a cliff-girt islet, then the echo is strong enough to be recorded and will appear to be a much smaller object only 50 miles away. As the plane approaches a point 100 miles away from the islet (and the signal apparently indicates an object closer than 50 miles away), the signal will disappear, because at 100 miles the echo returns in 1/250 of a second, when the receiver is blanked out for the next pulse. Thus the disappearing blip, instead of being a forewarned U-boat, was a second-time-round echo from an islet, a finding that the Radiation Lab expert and the crew proceeded to verify by flying over the piece of rock.

Many small islands appear in the Caribbean, so it was necessary to devise some means of distinguishing between true echoes from nearby targets and second-time-round echoes from more distant, false targets. That was easy, once the facts were clear; the radar set needed only a switch to change the repetition rate momentarily. This change would make no difference to the first-time-round echoes, but it would shift the apparent distance of a second-time-round echo. The switches were soon installed, and the fliers went back to using their radars. After the war, we found out that the Germans did try to equip their U boats with radar detectors, but the first installations were faulty and gave so many false alarms that the sub skippers gave up and refused to turn them on.

Our performance in this relatively trivial case demonstrated that our group could be of great value during the crucial period in the introduction of any new equipment, either ours or the enemy's. The actual users of new gear usually do not understand it at first, dislike it because it is unfamiliar, and quickly turn against it if it doesn't produce the promised results right away. In World War II, when radically new equipment frequently appeared, often accompanied by only sketchy instructions, it was imperative to have technically knowledgeable people close to the users, to amplify the instructions, to explain why the new gear should work better than the old, and to help fix minor malfunctions before the equipment got a bad name.

The disappearing-blip experience also demonstrated that we had been right in insisting that our field men return to the central group at the end of a six-month tour. The replacement man would go out primed with the latest developments, and the returned one would bring back to the equipment designers—at the Radiation Lab, for example—information as to how the gear could be improved. We learned, after the war, that neither Germany or Japan had operations research teams working with troops at the operational level. We had long suspected that this was the case, partly because of the amount of fumbling the enemy forces went through whenever new equipment, ours or theirs, was introduced.

Another study, begun under the pressure of events, and this time carried out in our Washington office, had to do with a new German weapon; it showed again the importance of quick reaction to change. Reports trickled through from the French underground that the Germans had developed an acoustic torpedo that would head for the sound of a ship's propellers. These self-guided weapons would be particularly effective against escort destroyers, which were heavily powered and thus noisy. It became imperative for our side to get hold of an actual torpedo or otherwise to find out about its characteristics, so as to develop countermeasures before the Germans learned to use the weapon effectively. Our group was in a good position to make such a study, for the underwater sound measurements from the MIT project organized in 1941 to counter the acoustic mine could be just as useful in countering an acoustic torpedo. However, we needed to know more about the properties of the torpedo: what range of frequencies did it respond to, and what was the nature of the linkage between its directional listening device and its steering mechanism?

By a lucky fluke, the Navy listening net learned at just that time that a submarine had come out from the French coast and was waiting in mid-Atlantic to join its mates, at a spot providentially close to one of our recently activated escort carrier task forces. This force, with its destroyers and ASW planes, found the sub on the surface, crippled it by air attack before it could dive, and boarded it from a destroyer before the crew could scuttle it. The boarding party found no acoustic torpedo aboard the U-boat, but they did find that the chief torpedo man in the German crew had watched an acoustic torpedo being demonstrated and had been told he would have some on his next cruise. Presumably the German submarine command was showing off its prospective weapons in order to bolster crew morale, which had begun to flag as our ASW tactics improved.

This torpedo man was brought to Washington for questioning. We were not allowed to see or talk to him, but we had daily conferences with the intelligence officer who carried on the questioning, suggesting questions for him to ask and getting a detailed account of the answers.

Invention

As far as I could tell, the prisoner was not threatened or maltreated; as we now are quite aware, persistent, friendly conversation is often all that is required.

The torpedo man unfortunately didn't have much to tell. In one of his refresher briefings in France, an acoustic torpedo had been wheeled in, its steering action had been demonstrated, and the nose cover had been removed to show the general layout of the acoustic gear. The lecturer had shown that noise above a certain intensity, coming from the right, produced a response of full right rudder; noise from the left snapped the rudder over to the left, and noise from ahead, or no noise, brought the rudder to neutral. All our prisoner had been able to see when the nose cover was removed were the approximate size and relative position of the microphones and the amplifier box. But that was enough to get us started.

We immediately recruited Edwin Uehling from the MIT underwater project, and we talked to the experts who were designing an acoustic torpedo for our own submarines. From the reported size of the microphone cases it was estimated that the torpedo responded to high-frequency sound, probably between 20,000 and 30,000 cyles per second. (After the war we learned that the response had been peaked at 26,000 cycles.) From the report on the steering demonstration and from the location of the microphones, we could make a good guess as to the kind of track the torpedo would follow as it chased a ship's propellers. And from the report that the torpedo was battery-powered we could estimate its speed.

Uehling began plotting sample tracks for different starting points of the torpedo and different speeds of the target ship. Then, assuming the existence of a decoy sound source being towed away from the ship at different distances and relative positions, he plotted more tracks to see when the torpedo would chase the decoy rather than the ship.

After two intensive weeks, his best estimate was that one noisemaker, towed well astern, would lure the torpedo away from the ship, unless the torpedo was released from nearly ahead of the ship. He could also report that our parallel-pipe noisemaker should make enough noise at the required frequency. This device, as we had found earlier, was easy to make and could be towed with little trouble. The British had also been working hard on the problem of countering the acoustic torpedo. They preferred their solution of an air hammer inside a metal drum—the same device we had tested on the range in Boston Harbor—and they felt that each ship would require two of them, towed by paravanes, one on each side.

We held to our findings, and our Navy went along. Soon our escort vessels had parallel-pipe decoys ready at the fantail, with orders to release one whenever the presence of a U-boat was suspected. The Germans had already begun to use the acoustic torpedoes, and one of our destroyers was damaged before it had been supplied with decoys. Our recommendations proved to be justified. We lost no more destroyers after the decoys came into use, and there were a number of reports of parallel pipes being blown out of the water by confused torpedoes. The more complicated British countermeasure didn't seem to work as well, or else it wasn't used often enough because it was hard to use. Several of their escort vessels were hit by acoustic torpedoes in the ensuing months.

We heard, after the war, that the U-boat crews were

Invention

terrified by the loud buzzing of the pipes. They called them "singing saws" and were sure they were some powerful, dangerous weapon. Because of our quick reaction, the Germans never achieved the success they expected from their new torpedo. As usual with any innovation that does not succeed quickly, the submariners came to distrust it and used it less and less.

Meantime our group continued to grow, in numbers and in variety of tasks. We sent John Pellam to North Africa, where he arranged a patrol of the Straits of Gibraltar using a magnetic submarine detector hung from a Navy blimp. That patrol caught two U-boats and discouraged others from entering the Mediterranean during the landings in Sicily. We sent Jay Steinhardt to Brazil, where he laid out air patrols between Africa and South America that caught most of a string of German ships bringing home needed rubber from Japan.

And we were finally asked to look into problems in the Pacific war, going beyond ASW to naval problems in general. Rinehart was assigned to the headquarters of our own submarine force at Pearl Harbor. He set up a duplicate of our Washington computer installation to analyze our burgeoning war against Japanese shipping. Our submarines used a radar set to spot their prey and to warn against Japanese ASW planes. Just as with our pilots in the Caribbean, Rinehart had to keep persuading the submariners not to turn the radar off, showing them that the Japanese did not yet have a radar listening device. After the war, a Japanese physicist, with tears in his eyes, told Compton that he had urged the development of electromagnetic detectors (i.e., radar) but had been turned down.

Fergie Brown (he was the second Arthur Brown in the group, so he had suggested, "just call me Fergie") went out to Kaneohe, Hawaii, to assist Naval Air Operations. And after his tour in Brazil, Steinhardt was sent out to the headquarters of the Seventh Fleet, assigned to Gen. Douglas MacArthur, first in Brisbane, later in Hollandia. These men all did highly useful work, and the backup analyses done in the Washington office broadened their scope and occupied more analysts. At the same time, my job edged further and further from the exciting details of the problems to the management of people: to arranging the installation of a new man in the field, to soothing the feelings of an analyst who had not been chosen to go, to persuading a group member to drop an interesting calculation to take up a higher-priority problem, and then to calming the irate admiral for whom the abandoned calculation was being made. More and more I seemed to be dealing in words rather than facts. I did manage to assign a few men, such as the mathematician Bernard Koopman, to major, long-range studies, such as the consolidation of search theory, and to protect them from all but the most urgent interruptions. And then I spent my spare time envying them.

The only times I was able to get away from pushing words and people around was to attend some NDRC conference or to go to the West Coast to see the Cal Tech people about rockets, ASW or other. The plane trips from Washington to Los Angeles were usually made overnight in DC-3's so heavily loaded they had to refuel three times just getting across Texas: at Dallas, at Big Spring, and then at El Paso, just about when it was getting light. I had a priority, or I wouldn't have got on a transcontinental plane at all. But I didn't have as high a priority as did the shuttle pilots, who flew planes from the West Coast factories to the Texas air-training stations and then took commercial planes back to the Coast. I never knew whether I would finish out the long night in the DC-3 or on the bench at the Dallas airport.

Most of the time I stayed in Washington, keeping things running. By the end of the war, our group, then known as ORG, had grown to include nearly a hundred analysts, with ASWORG being but one of the subgroups. No members of the group were killed during the war, although many of them got close to the action. Rinehart went on a month's cruise to the Japan Sea in one of our submarines, and Kip was on a ship hit by a kamikaze plane off Okinawa. Maurice Bell was half-scalped when, after a routine ASW patrol flight, the plane he was in landed in an orchard instead of on the runway, but he returned to work after a month.

After the German surrender, a few senior group members went to Europe to verify our analyses by checking them against the German records. M. S. Livingston, for example, spent some time interviewing Admiral Doenitz, who had commanded the U-boats throughout the war. The rest of the group turned to the continuing struggle in the Pacific, helping Rinehart with our own submarine force, assisting Steinhardt, first in New Guinea and then in the Philippines, or working with the landing forces at Iwo Jima and then at Okinawa.

As the war neared its end, most of the group became increasingly eager to return to their peacetime occupations. But many of us felt that, before we disbanded, our analyses and achievements should be recorded, to be available in organized form in case like problems arose in the future. Beyond this, several of us in the Washington office who had become familiar with the military command structure felt that we could contribute to plans for what we saw as the inevitable reorganization of the armed services, so that they could operate more effectively in times of rapid technical progress and so that they could be more effectively controlled by the civilian government. The Navy agreed, and a smaller edition of the group was kept on, under Steinhardt's direction. In modified form it still exists.

Many of us were also convinced that the methods of analysis the group had developed—that is, operations research—could be applied equally well to nonmilitary operations, such as industrial production and distribution or perhaps even running a city. A detailed discussion of how the techniques had been applied successfully to military operations would, we hoped, persuade industrialists and public administrators to try out the techniques—and the analysts. We hoped that this kind of scientific study of the cooperative actions of men and machines, begun in wartime, might be applied to more humane activities.

In addition to sending group members to Europe to learn where we had been right and where we had been wrong in our assumptions about German plans and actions, we arranged interviews with our own senior naval officers

Invention

to determine how useful the work of the group had actually been to them. Art Kip and I were on such a mission in the second week of August 1945. We were in the temporary office of Adm. Willis Lee, in the naval base in Casco Bay on the coast of Maine. Admiral Lee had just returned from the Pacific. He had been in much of the fleet action there, from the battles off Guadalcanal in 1942 to the latest melee at Okinawa. We had become acquainted with him during his occasional tours of duty in Washington; he had been interested in our studies of radar fire-control accuracy, and we had been impressed by his technical knowledge and his quantitative approach to tactical problems.

Admiral Lee was a small man—many admirals are—usually quiet, with a more scholarly than military appearance. This time he seemed worn out, physically and emotionally, by his recent responsibilities. He would talk rapidly and almost compulsively for a while and then would fall silent, looking at the floor. He seemed glad to talk to us and recalled our earlier conversations with him. We asked him a few leading questions and then let him set his own pace, remaining silent when he paused.

In one of the periods of silence I heard a radio in the outer office interrupt its regular program for a news bulletin. The radio was not loud enough for me to hear the announcer very well, but I did get fragments—about a new kind of bomb ... atomic energy ... dropped on a Japanese city ... great destruction ... And then Admiral Lee began to speak again and I couldn't hear any more. We spent a further hour, I torn between my desire to hear the admiral out and my burning curiosity to hear the rest of the news announcement. As soon as we could do so politely, we excused ourselves and returned, on the admiral's gig and by official car, to Portland, where we could pick up a newspaper extra. That was our last talk with Admiral Lee. He died suddenly less than a month later.

The announcement, of course, was the first one about the A-bomb. It came as a complete surprise to most of the Navy officers we knew, as well as to most civilians. George Kimball and I had had some inkling of the development of the bomb, although we were not cleared to receive information. George had been in a part of the atomic project before he came to the O/R group, and many of our friends had vanished from their usual places to go to Los Alamos. Knowing the fields of specialty of those friends and reading their faces as they evaded supposedly harmless questions had given us a pretty good idea of the nature and progress of their work. Late that spring their manner suggested that they had succeeded. Naturally we said nothing to anyone else, but we had been alert for news of the results. Kip was surprised at the amount of detail I could fill in between the bare statements in that first public announcement.

By the time I got back to Washington, Abe Olshen, our office manager, had contrived to get his hands on one of the few initial mimeographed copies of the report written by my Princeton friend Harry Smyth that gave the few details of the A-bomb and its development that were deemed safe to release at the time. Kimball and I spent a long evening going over it, filling in the lacunae with our knowledge of nuclear physics and of air and sea warfare. The next day we gave a presentation to the ORG of the properties of the bomb and its implications for the Navy of the future. This presentation was attended by some of the staff officers working with us. Their reactions were such that we were requested to repeat the performance for Admiral King and his staff.

An indication of the tightness of the security control achieved up to then by the Manhattan Project was that not even Admiral King had been told enough about the new weapon for him to estimate its size or its lethality. A few naval officers had been assigned to the project, such as Comdr. William S. Parsons (later Admiral Parsons), who was on the plane that dropped the bomb on Hiroshima. But these officers had been sealed off from the rest of the Navy, so the facts and potentialities we presented that afternoon were completely new to most of the top echelon of our Navy. Many questions were asked us about the chances of survival of a carrier or a battleship and about the possible effects of an underwater blast.

Admiral King was impressed enough with our talks to mention them to Secretary of the Navy James Forrestal. So the next evening, we gave another command performance before the Senate's Naval Affairs Committee. Kimball and I spent some time with Forrestal after this session, explaining what our group had been doing and how it happened that we could put together the sort of talk we had just given. This was our first contact with Forrestal. It was to bear fruit later.

Once VJ day was past, nearly all the group were eager to be gone, and it took all the persuasion George and I could muster to keep a skeleton force long enough to finish our final report. The report shaped itself into three volumes: the first on the methods of operations research, the second on the history of the antisubmarine war, and the third on search theory and its application to naval problems of search and screening. The division was made in the hope that we could quickly get the first volume declassified, so as to bring the name and the methodology of operations research before the general public and, we hoped, arouse interest in its application to peacetime problems. To this end we had avoided classified examples and had written the book in the form of a text. However, when the three volumes first came out, they were all classified Confidential. (We had to wait six years for the bureaucratic wheels to turn enough to release the first volume.)

Finally, in the early fall of 1945, I could return to Boston and begin to restructure my life into its old patterns—I hoped. But could I still do research and would I still relish teaching?

8
Initiation

Some time in August 1945, a group of us from the MIT physics department, including Slater, Stratton and Allis, were sitting on the lawn in front of Slater's summer cottage in South Newfane, Vermont. We could look out south across the valley toward Massachusetts. The occasion was a reunion and a homecoming. We had all been away from the affairs of the department. Stratton had been with Bowles and Shockley in the high-level consulting group of the Air Force; Allis had been a lieutenant colonel in the Army's Office of Scientific Research and Development; Slater had been in New York with the Bell Telephone Laboratories, designing more powerful radar magnetrons in collaboration with Jim Fisk, whose doctoral thesis I had supervised; and I had been in Washington most of the time (and was, in fact, still finishing up the ORG project there). Each of us had learned how to manage big projects and had participated in important policy decisions. Now we were gathered to plan the future of physics at MIT.

We were glad to be back, but none of us was sure it was or could be a true return. None of us was the same as he had been in 1939, and we suspected that our lives could not be a return to prewar days. Whether we liked it or not, physics and physicists had transformed warfare and would inevitably have a profound influence on the peacetime progress of the nation. We expected that many of us would be involved in channeling and monitoring this influence. On that pleasant summer afternoon, I, at least, was not aware of the amount of energy and time this kind of public service would require, nor of the number of instrumentalities that would have to be formed. Planning the future of physics at MIT was the task at hand; more global ones would come later.

In the fifteen years since Karl Compton had taken over as its president, the Institute had grown from a respected engineering school to a world-renowned scientific university. By the end of the war, well over half its space, personnel, and budget were devoted to research and development for the war effort. It now was necessary to transform the Institute back into an educational establishment, with no reduction in its preeminence.

Physicists had been the wonder boys of the war effort; they were likely to be expected to produce other marvels. Ambitious young men and women who could pass algebra would want to become physicists. Our department would have to expand; many more physicists would have to be educated for research, for teaching the next generation, and for industrial and governmental posts that would require knowledge of physics and physicists. The Radiation Laboratory, which had developed the wartime radar equipment, could not be scrapped. Most of its know-how and instrumentation would have to be used to teach the next generation to solve peacetime problems.

Stratton was already scheduled to take over the transformation of the Radiation Laboratory into the Research Laboratory of Electronics. There was some talk, that afternoon, of that operation, as well as of the formation of an Acoustics Laboratory, a minor development but one of interest to me. But most of the discussion was focused on the field that would need the greatest expansion, nuclear physics. We at MIT had competence in electronics, in quantum physics, in acoustics, and in basic theoretical physics, and we knew the latest applications in these fields. We had some homegrown experts in nuclear physics, notably Herman Feshbach. But if the department was to have adequate capability in that awesome new subject, we would have to import more such experts, and the obvious sources for them were the Manhattan District laboratories. We would have to recruit people from Los Alamos and Oak Ridge.

With us that afternoon was Jerrold Zacharias, who had agreed to head a proposed Laboratory of Nuclear Science at MIT, planned to equal the Electronics Laboratory in size and competence. Zach had worked at Columbia under I. I. Rabi before the war, had come to the Radiation Laboratory when it was started at MIT in 1939, and had left to go to Los Alamos in 1945, where he had directed the engineering division. He wriggled with enthusiasm as he listed the stars he expected to recruit: Victor Weisskopf for nuclear theory (I had met him in Germany many years before), Bruno Rossi for cosmic rays, and a number of younger men who had shown their worth at Los Alamos, including Bernard Feld, Martin Deutsch, David Frisch, and others. Expensive equipment would have to be built or purchased, but this was not the insuperable barrier it had been before the war. Government agencies, particularly the military and the Manhattan District (soon to be transformed into the Atomic Energy Commission) were ready and anxious to support university research.

Before I had left Washington I had had long talks with

Capt. Robert Conrad, of the newly formed Office of Naval Research (ONR). He was sure that the country had at last learned its lesson, that basic research in the universities would have to be supported by the federal government. The question was, How would the support come? I had my own ideas about that; I felt strongly that the armed services should not provide the major support. The military did, of course, need the products of university research, such as new ideas about potential weaponry and a continuing flow of trained men to become technical officers and to staff military development laboratories. But the output of university research, I believed, was even more important for the peacetime development of the country, and I was afraid that if the military controlled most of the funds, the directions of research would inevitably be bent away from peacetime goals.

Bob Conrad agreed with my worries; he had had a peacetime indoctrination in the importance of civilian control. But, he pointed out, a new civilian agency would have to be set up to administer the funds and this would take time, Congress being what it was. What was to be done in the intervening three to five years? The universities needed immediate support; without it, postwar research and scientific education would never get started. The only governmental agencies that had learned how to support university research were the armed services. Conrad was writing the charter of the ONR so it could support basic research without being too specific or restrictive in its guidance.

I had to admit he was right in his pragmatism, and I had a private fear that a civilian agency might not get as much money for basic research as the military could pry out of Congress. But I was still bothered. If the military began the support of university peacetime research, would this support make it harder for a civilian agency to be established? We spent a good many hours discussing these questions, and I left Washington much relieved that the Navy had chosen someone as thoughtful as Conrad to run its program of support for university research. Unfortunately, two years later Captain Conrad developed leukemia and had to retire from the Navy. He came to work for me at Brookhaven during his last year of life.

In the field of nuclear physics, the prospects were not as promising. When I got back to Cambridge I began to hear about the struggle that had been taking place to free nuclear research from the clutches of the completely militarized Manhattan District. As my friends came back from Los Alamos and Oak Ridge, I learned about the extreme security restrictions that had been applied to the nuclear project. These shocked me. Of course it had been necessary to keep secret any indication of success in the development of the atomic bomb, as well as the details of the bomb's construction. But our Navy Operations Research Group had had to preserve secrecy about planned naval operations and weapons, as well as the fact that we had broken the enemy's code, and we had not been bullied and manipulated in order to keep us quiet. It was not the fact that secrecy had to be maintained; it was the way it was done that shocked me. The more I heard about it, the more thankful I became that I had never had to work under as cynical an administrator as Maj. Gen. Leslie R. Groves.

The results of this manipulation were, of course, an explosion of feeling that became public after VJ day and an abiding mistrust of any form of military control of research by many who had worked on the bomb project. It was an attitude I agreed with in general, but my war experience had not made me emotional about it. Still, emotion fired the energies of some remarkably intelligent individuals, who proceeded to show what supposedly unworldly scientists can do if they put their minds and emotions to work. A relative few of them fought the military and military-minded congressmen to a standstill and persuaded Congress to create an Atomic Energy Commission under civilian control.

By the end of September 1945, most of the prewar faculty were back at MIT, teaching and laying out plans for research. I had a chance to work again with Herman Feshbach, who had been teaching the Methods of Theoretical Research course while I was away. We started outlining the material we would include in the text we planned to write for the course. The more we talked, the more material we found we had to include. Despite the projected length of the book, we saw a need for a text that would collect in one place the various advances in methods that had been developed for theoretical physics during the preceding two decades.

All of us in the department started to learn how to raise money for research. Most of us became members of various advisory committees of the armed services. My first appointments were to the Naval Research Advisory Committee and the National Research Council's Undersea Warfare Committee. Attendance at meetings of such committees gave us a chance to maintain personal contacts with the various agencies and to sniff out new sources of funds to support the projects we were interested in. Put thus baldly, our efforts sound like collusion in the expenditure of government moneys, but I must confess I still do not see any other efficient way in which the changeover from wartime to peacetime research could have been achieved.

During the war, civilian scientists had run much of the military research and development, through NDRC. After VJ day all these activities were turned over to the military, and it was only natural that the armed services would ask for help in planning their continuing support of research. By 1945, the armed services had become converted to the thesis that scientists could help in the practical application of scientific discoveries, and the services wanted to continue to get advice from those carrying out the research.

In turn, we scientists wanted to keep in touch with the governmental agencies concerned with the application of research discoveries, whether for war or peace. During the war, many of us had learned that the result of a scientific discovery was not just an increase in knowledge but also an increased potentiality for modifying, for good or for bad, some aspect of human activity. Many of us began to consider—and sometimes to worry about—these potentialities; and we found it useful to talk over the implications of our research with our friends in Washington. Our talks with our military friends were not just about new weapons. Many military men were just as anxious as we were to find peacetime applications for the research that they supported.

My first task in fund-raising, to get support for the new Acoustics Laboratory, was not difficult. Dick Bolt had returned from an NDRC job in England, Leo Beranek had decided to come to MIT from his Harvard project, and a number of the younger members of the MIT underwater-sound project wanted to stay on, some of them to get the degree they had long postponed. The Navy wanted to continue some underwater acoustics research and also was willing to support a return to the room acoustics work we had shelved in 1939. Bolt and I had written a review of room acoustics for the *Reviews of Modern Physics* in spare moments during 1944, and I had spent weekends in 1945 revising my book *Vibration and Sound*. So we were agreed as to how the laboratory should begin. Bolt and Beranek were to put through the detailed proposals for funds; all I needed to do was to suggest whom to see and what key phrases, inserted in the proposal, would be most persuasive.

My own plans for teaching and research shortly had to be put aside again, as the more urgent plans for research in nuclear physics unfolded. Nuclear physics was deservedly the glamour field, just as quantum mechanics had been in my youth. Here one could have the excitement of exploring the entirely new realm of the atomic nucleus, as well as the fascination of its promise of vast powers. If an atomic bomb could result from the fragmentary discoveries of the thirties, what other potentialities might come to light with further research? Whether the applications of the new knowledge turned out to be benign or dangerous, we had to seek the knowledge.

To probe more deeply the structure of the nucleus we had to build more powerful accelerators to break apart the nucleus and so to permit the study of its fragments. To investigate the possible peacetime applications of nuclear fission, we had to build research reactors. Both accelerators and reactors were huge and costly pieces of equip-

Initiation

ment. No single university could possibly afford to build one, nor could a single university keep one busy. Either a group of universities had to band together to build and operate such equipment, or academic scientists would have to depend on governmental laboratories such as Oak Ridge and Los Alamos. None of my friends who had worked for General Groves could look foward to that latter alternative with much pleasure or assurance. The initiative came from Dean George Pegram of Columbia, acting at the urging of his colleagues Rabi and Norman Ramsey. He called together influential people from Harvard, Princeton, Cornell, and other major universities, in the Northeast, among them Zacharias, who was by then at MIT. By the time Zach approached me, in the spring of 1946, plans had been laid for a consortium of nine universities, to be called Associated Universities, Inc. (AUI). This consortium was to contract with the government to run a large nuclear research laboratory, to be located somewhere in the Northeast, whose primary task would be to build and operate a nuclear reactor for use by academic research scientists. To my astonishment, Zach asked me whether I would be the director of the laboratory.

My first reaction was negative. To accept would mean reentering the political-administrative arena, in the region of greatest political pressure, just when I was beginning to relax from the wartime stress. I had not kept up on developments in nuclear physics; my last research contact had been to supervise Herman Fesbach's 1941 Ph.D. thesis on the collision between the proton and the deuteron. Furthermore, I had not participated in the bomb development.

Zach's rebuttal was that the director's job would be to

pull everything together and to get along with Washington, and I had had plenty of experience of that sort. I could get the technical advice I needed from the many experts on the faculties of the participating universities. AUI would have a part-time president and a full-time vice-president and would be guided by a board of trustees, two from each university, one in administration and one in research. My job would be to get everyone moving in the same direction. Zach kept saying that the wartime effort has been an engineering triumph, that nuclear science had made no major advance since the discovery of nuclear fission in 1939, and that it was imperative for us to recover our pace in basic nuclear research.

It was a tough decision. This time I would have to move the family to a place near the laboratory and would consequently subject the family to many of the stresses that they had avoided during the war. Conrad was at a stage when a move might have adverse effects. Annabelle was willing but certainly not enthusiastic. In spite of this, I began to feel I should take it on; and Zach and other friends kept saying that there was no one else available.

As I was deciding whether or not to take on the job, I was taken by Zach and others to talk with General Groves; this visit almost made me decide to back out. Groves was still in charge of the giant Manhattan District establishment. The newly authorized civilian Atomic Energy Commission would not take control until the end of the year, and it would not do for the AUI project to wait that long. In any event, the AEC would take over the Manhattan District personnel.

The meeting was characterized by muted antagonism on both sides. Groves tried to be gracious, and his con-

Initiation

tempt for the impractical scientist came to the surface only a few times. I kept as quiet as possible, just answering direct questions and letting the others do most of the talking. The result seemed to be inconclusive, and I was surprised to hear a few weeks later that Groves had acquiesced in my appointment and had authorized a grant to get AUI started. I suddenly realized I was committed. I consoled myself by remembering that it had been agreed I would not have to stay for long. To me, the building of a new organization would be the best part.

By that time, the location of the laboratory had been agreed on. A committee of AUI had surveyed a number of possible sites and had chosen a decommissioned army camp on Long Island as the best in regard to central location, extent of grounds, and number of usable buildings. Camp Upton was near the town of Yaphank, midway between the south and north shores of Long Island, sixty miles from Manhattan, and well away from any populated areas. We could not move into the site until the spring of 1947, nine months hence; in the meantime we could make a start in offices allotted to us at Columbia.

Nine months was none too long a time to get organized. We had to recruit a staff, to decide what facilities we wanted to build, to determine our organizational structure, and to define the interface between our administration and that of the Manhattan District (and later the AEC) bureaucracy. I felt strongly that this last task was supremely important. We had to think through what we should take responsibility for and how we should exercise this responsibility, before the governmental bureaucracy decided for us. If, for example, we let the government provide the guards at the site, red tape could neutralize our aim of making it easy for university scientists to come and go. I wanted to have a say in these matters, within the rules of security, of course. I wanted to keep the initiative and to leave only a veto power with our governmental overseers.

To keep the initiative, I needed some good people quickly. I could find some of the scientific staff at the associated universities. Norman Ramsey came from Columbia to head the physics department; he set about recruiting the rest of his staff. I soon found heads for the biology and chemistry departments, but it took several years to find a head for the medical department and to formulate a program of medical research. All these departments were to use the big equipment, the reactors and the accelerators, which were also to be the attraction for the more temporary attendance of scientists from the participating institutions.

We were lucky in getting good men to head the groups that were to plan and eventually to operate these big instruments. Stan Livingston had worked under Ernest Lawrence at Berkeley in building the first cyclotron; he had worked for me in the Navy ORG during the war. He started visiting campuses to find out what sorts of accelerators would be most useful. Lyle Borst came to us from Oak Ridge, where he had played an important role in designing some of the wartime reactors. He was eager to get started on the first peacetime atomic pile.

Before the technical staff could get very far, I had to collect an administrative staff, to take and hold the initiative I desired. My very first recruit was a woman, one whom I had known in Princeton when she was Johnny von Neumann's wife. Since then she had been divorced, then married to J. B. H. Kuper, whom I also wanted to recruit as head of health physics. Mariette Kuper was exceptionally talented—sharp-faced and sharp-eyed, exuberant, voluble in English or Hungarian or her usual mixture of both, but great-hearted and deeply understanding, with an unerring intuition as to the right way to handle people. She had spent part of the war training radar technicians, using methods she devised herself, and she was enthusiastic about our plans. The first job I gave her was to get herself hired and cleared, which she accomplished smoothly and miraculously in less than a week. She became my one-person department of human relations, of official hospitality, and of odd jobs that no one else knew how to do.

Another assistant was Lincoln Thiesmyer, whom I had come to know during the war when he was one of Compton's administrative assistants in NDRC. He knew the Washington scene and agreed on the need for keeping the initiative with us. Personnel and site planning were two crucial areas. I wanted heads of these two departments with enough authority and experience that they would not be browbeaten by the Manhattan District people. We found them. We got Lawrence Swart, who lived out near Camp Upton, who knew personnel and security procedures, and who recruited a former FBI agent who could head our security force. The whole staff worked out methods of clearance, guard procedures, and secrecy control that would satisfy the rigorous Manhattan District requirements and yet be as unobtrusive as possible. And I persuaded Georges Peter, a member of the Harvard School of Architecture who had been with the Radiation Laboratory during the war, to take care of buildings and grounds and to plan the new buildings and site changes.

For a while, we all spent time arguing about how we could admit an uncleared visitor from one of the participating universities and yet keep him from the secret parts of the laboratory, without making him feel like a second-class citizen. More than half of our activity would be as open as any university research; none of the accelerators would be classified, for example. On the other hand, the workings of a nuclear reactor were then highly classified, and part of our library would consist of secret documents. To avoid the easy and destructive solution of requiring clearance for everyone using the equipment, Peter laid out the site and planned the reactor building so that classified activities were segregated and well guarded; the rest of the site then needed no more patrolling than is required in any large plant. Visitors to the unclassified area would not have to be catechized at the gate, but would be directed to the Visitors Office, where they could be unobtrusively guided to the person they came to see. By the time we moved to the site, our procedures were organized and were largely acceptable to our governmental overseers. We even designed the reactor building so that one side of the reactor was open for workers without security clearances.

The next step was to test whether all our full-time personnel would have to have the full security investigation required of those dealing in atomic secrets. Of course, everyone involved in the reactor design and operation would have to be subjected to what was coming to be called "Q clearance," with its delays and multilevel screening. Our reactor planners all came from Manhattan District laboratories and already were cleared. The people to design

Initiation

and operate the accelerators posed a different problem. The work they would be doing would be unclassified, and many of those we wanted did not have a Q clearance. We worked out a procedure for those cases: the prospect was asked to choose between waiting the usual three or four months to get a Q clearance (with a not insignificant chance of being turned down) or to come without clearance, understanding that he would be barred from some areas and activities.

The case that proved the rule was that of John and Hildred Blewett. Man and wife, they had worked for General Electric at Schenectady on problems closely related to accelerator design. But they had openly sided with the union in a turbulent strike at the GE laboratories, and they and their employer had become estranged. We felt sure that the two could make major contributions to the accelerator design; they were anxious to come and did not want to have anything to do with classified activities. This was the first time we proposed to hire senior staff members without having them cleared, and we had to fight through a complete bureaucratic defense in depth to get them.

We had already weakened the defense on the question of hiring less senior personnel without clearance, as long as they were excluded from sensitive areas. But the AEC had the final approval of salaries above a certain level, although up to that point the agency had always approved salaries that had been approved by the AUI trustees. Things were brought to a head when the Blewetts left Schenectady in a minor puff of publicity that alerted all sides. They arrived on Long Island expecting to go on the payroll the next day.

For the first time, some of the administrative trustees hestiated to approve my recommended appointments. And then, when the trustees were eventually convinced and approved the Blewetts' salaries, the AEC bureaucracy began to raise questions about hiring a man and wife, about whether they merited the proposed salaries, but never directly about their lack of clearance. It took more than a month of argument and several trips to Washington on my part to get approval. By then the Blewetts were living on charity. They soon showed their worth; they were responsible for a number of original and successful design concepts in our first big accelerator.

We didn't win them all. We lost the fight over Ronald Gurney, the tall, silent Englishman I had worked with at Princeton. He had worked in England for the British atomic energy effort and, when the Manhattan District took over, he had been sent over to work with our people. At the end of the war his job was quickly terminated, Groves's people had never liked having Britishers learning "our" secrets. By the time Gurney had heard about our new laboratory and applied for a job in the physics department, the Fuchs spy case had broken, and Congressman J. Parnell Thomas was ranting about traitors in our midst. It was impossible to get a Q clearance for Gurney, although he had been cleared during the war and no derogatory information about him was ever found. This time I could not persuade my board to back me in hiring Gurney without clearance, so I had the hard task of telling Ronald that he could not work for the laboratory, even though I was sure he would have been a major addition to the staff.

Gurney spent the next several years subsisting on odd jobs, trying to get a clearance so he could work on the

Initiation

research he wanted to do, barred from the subject he had helped to start—he and Condon had been the first to explain radioactivity in terms of quantum mechanics. He died in 1950, a victim, I feel certain, of the treatment he had received, a sacrifice to our paranoiac fear of losing "the secret."

But this is getting ahead of my story. My family moved to Bayport, Long Island, in October 1946 and began adjusting to the differences between Massachusetts and New York in schools and in local customs. The fact that Annabelle had to tend to the complex details of selling our pleasant house in Belmont, of finding a livable one on outer Long Island, of supervising the moving and of adjusting the children to the change hardly endeared the transfer to her.

During the following winter I commuted in to Columbia, via the Long Island Railroad and subway, more than an hour each way—even when the Long Island was on time. It did not convert me to commuting.

Most of the initial staff spent extra time talking to the townspeople living near the future laboratory, explaining what we were planning, the safety precautions we were going to take, and the fact that we were not going to be making atomic bombs. I talked to the Kiwanis and Lions clubs, held press conferences, and had radio interviews. The effort paid off; we had little of the local antagonism that many of the other labs were suffering under. By the time the reactor was working, many of the local people were working at the laboratory, and we had become part of the community. Long before then we had finally agreed on a name for the laboratory. We settled on Brookhaven, the name of the township. We could not use it for our post office address, however, so for this we settled on Upton, New York, thus preserving the camp's original name for the post office.

At the time our work was getting under way, many people in this country were convinced that if we let out the secret of the bomb, the country would be in mortal danger. On the other hand, we were sure—and events were to show we were right—that our preeminence was really in our ability to forge ahead. Any country's scientists, if they tried hard enough, could learn existing secrets by themselves. But if we kept learning more about nuclear science and engineering, we could keep ahead of the others. We knew that progress in science is always hamstrung by secrecy restrictions. Our laboratory was going to learn more about the nucleus and how it could be used for people's welfare. The more openly our findings were published, the more quickly other scientists in other laboratories could build on these findings to advance still further. Progress in pure science is best made by the open participation of all the scientists who are interested, not by a few shut up in a classified laboratory. Weapons design should, of course, be kept secret, but we were building new knowledge, not weapons.

The battle to open up free exchange of scientific ideas had to be fought on two fronts. We had to work through the bureaucracy and also take our case directly to the public, the final arbiter. As part of our tactics of keeping the initiative, we proposed a new publication policy for Brookhaven. A laboratory committee would have the responsibility for determining whether a report or a scientific paper issued from the laboratory should be published openly or not. In all the other AEC laboratories, *every* report or paper had to be submitted to Washington, to be closely scrutinized at great length to determine whether any of it could be published openly.

We argued that the rules appropriate for weapons laboratories were not appropriate for Brookhaven and that we would fail to become a cooperative center for university scientists if every visitor found his papers subjected to censorship and long delays. We pointed out that the great majority of the reports issuing from the laboratory would be unclassified, that our staff would know better what should be kept secret than would someone not connected with the research, and also that it would be psychologically better to give the laboratory the responsibility for security in this as in other matters.

We got little support from other AEC laboratories in this fight. During the war, all of their reports had been secret, and, even after the war, most of them still had to be classified, since they were related to weapons design or production. I was disappointed that Ernest Lawrence, head of the Berkeley laboratory, did not support us. For a while it looked as though we were going to lose this battle. I had to carry the case clear up to the Commission itself to get approval. The decision turned out to be the right one; there have never been charges that the laboratory published something that should not have been published and there have been no long delays in approving publication of papers that had no "secrets."

During this difficult time I first came in contact with an organization devoted to the broader effort to enlighten the citizen about scientific matters. The group of scientists that had fought to put the AEC under civilian control had realized that public understanding of the dangers and potentialities of atomic power and weaponry was urgently needed if this country was to act realistically and perceptively on the international scene. The group incorporated itself as the Emergency Committee of Atomic Scientists. It persuaded Einstein to be its president (and thus it was usually known as the Einstein Committee) and had Joseph Schaffner as executive director and Michael Straight of the *New Republic* as secretary. The board of trustees included Harold Urey, Leo Szilard, Hans Bethe, T. R. Hogness, Linus Pauling, and Victor Weisskopf. They held press conferences and meetings to raise money. A speakers' bureau was organized, and some of the funds collected went to support a new publication, the *Bulletin of the Atomic Scientists,* and to help organize a new society, the Federation of Atomic Scientists, both to be devoted to public education.

All these activities were well under way when I came to Brookhaven. Because of my proximity to the committee office in New York, and perhaps also because of my public appearances on behalf of the laboratory, I was asked to be a member of the board of trustees. I was glad to accept. It was a stimulating group. Urey, whom I had known since Princeton days, was voluble, almost stuttering in his effort to keep up with his spate of ideas; Bethe was quiet, his precise, slightly Germanic sentences always expressing a well-marshaled thought; Szilard was subtle both in speech and ideas, yet more practical and more dedicated than the rest of us. And there was Einstein, who very seldom said anything, but who seemed to look right inside each of us as we argued. His smile or slight frown, at the end of the session, evidenced his understanding of what had gone on and indicated the degree of his approval.

The activities of the Emergency Committee seemed discreditable to many of our colleagues. Some of them felt that a scientist somehow demeaned himself when he talked to other than his specialist colleagues. At the end of the war, scientists, particularly nuclear physicists, suddenly found themselves held in general awe; their pronouncements on any subject were considered important. Many were quite satisfied to take advantage of the esteem without honoring the related responsibility of explaining how science advances and why it can be important, not only for weapons but also for the general welfare.

There was some suspicion that scientists were pleading their own case if they took part in public debate. It is, of course, not easy to separate factual statements, designed to enlighten the layman, from arguments favoring some particular policy. The scientist must play a double role—as a specialist, to explain to others the relevant facts and, as a citizen, to take part in reaching public policy decisions. He has both a right and a duty to act both parts, but he needs, each time, to make clear to himself and to his audience which role he is playing.

The implications of the atomic nucleus were mysteries to the layman. Contrasting with the monstrous terror of nuclear warfare and the stealthy dangers of radioactive contamination were the potentiality of nuclear energy and the promise of major medical usefulness. The nation could make safe yet gainful choices in the years ahead only if it was given a balanced picture of both hazards and benefits and of the need for further research. The Einstein Committee's self-assigned task was to begin the education, through its own actions and also through those of the permanent instrumentalities, the *Bulletin* and the Federation.

By the middle of 1948, many of these tasks were coming to completion. The AEC had been established, independent of the military forces. People were beginning to understand the dilemmas of nuclear power. The Federation of Atomic Scientists was well established on many campuses. It was agreed that the *Bulletin of the Atomic Scientists,* so ably edited by Eugene Rabinowitz, needed further support, but beyond this it was not clear where the committee should go next. Some of the trustees wanted to continue raising money and assisting Congress on policy problems related to nuclear weapons. But it is the nature of money-raising that each new appeal must be supported by a new pronouncement capable of arousing new interest. It was getting more and more difficult to get agreement on new pronouncements. Some of the group wanted to continue to emphasize the dangers of the bomb; others wanted to address more political issues.

By the spring of 1948, most of the trustees were ready to agree that the committee had nearly finished its job. We found an able nurse for the *Bulletin* in the new publisher of the *Scientific American,* Gerard Pile, who was willing to advise Rabinowitz on the organizational and financial side of publishing. With that the committee wound up its affairs. In retrospect, I feel that the disbanding was healthy. It made way for newer organizations with fresh appeal, but we all found it a wrench to dissolve a successful organization.

The more active members went about setting up successor organizations. Leo Szilard, for example, spent

Initiation

several years in working out an appropriate form and function for the Council for a Livable World, which is still active and a continuing demonstration that not all scientists are politically naive.

But I have strayed from the story of Brookhaven. Public education was only part of the job of getting the laboratory started. The decisions about the big machines presented a number of difficulties, both expected and unexpected. Not many disagreements about the reactor arose among the scientists who planned to use it. Neither they nor we wished to design a new type; what was wanted was a plentiful source of neutrons so that their physical, chemical, and medical effects could be studied. The chief design innovations were to make the system safer for the experimenters and entirely safe for the surrounding population.

We chose an air-cooled, graphite-moderated design, since there had been more experience with this type than with any other. The coolant air was to be recycled, with none of it getting out; it was kept at a pressure lower than atmospheric, so any leak would suck air in rather than blowing radioactivity out. This internal air would be cooled through a heat exchanger, and the coolant air would be exhausted through a tall chimney for extra safety. Multiple devices would ensure a complete shutdown in case of a leak from the fuel rods. Each safety device was to be repeatedly tested under simulated emergency conditions.

The difficulties in getting the reactor construction approved were not technical; they had to do with the choice of the company that would carry out the detailed design and construction. The AEC people wanted to use one of the companies that had constructed the Oak Ridge reactors. Our expert, Lyle Borst, did not have quite so high an opinion of the company that the AEC recommended. Borst wanted to spread the know-how of reactor construction to other companies, and he suggested that we choose a company that had a high reputation for design in some other field, such as naval ship building, and that had solved analogous problems of safety amid complex machinery.

We lost that argument. The AEC appointed the company their people knew. As Borst had foreseen, the company's excellent wartime staff had largely moved on, and few of the engineers they assigned to the job had worked on reactors before. Most of their mistakes were caught by Borst's group, and our only final complaint was that it took a year longer and cost about five million dollars more than it should have. The reactor remained in use until 1969; it operated satisfactorily and never had a leak or other accident.

When we came to build the accelerator, the AEC hierarchy, having had no experience with this equipment, had no companies to recommend. They let us have our way, except, of course, for keeping a close watch on the budget. Our laboratory staff was to do all the design work, which meant we had to hire a large engineering staff. We could borrow some members temporarily from the participating universities, and there were others, like the Blewetts, who wanted to gain experience in the design of accelerators and who would be willing to go elsewhere later to design other, bigger machines.

Our arguments here were with the university physicists. Some wanted high intensity; others, high energy.

Initiation

Some wanted protons as the accelerated particles; others would be satisfied with electrons. Still others wanted the design group to try radically new designs, in the hope that a higher voltage would be attained with less material and thus less cost. Both Stan Livingston and I felt that we should stick to a fairly conservative design, one that we could be sure would be completed fairly quickly and would produce the planned energies and intensities. We feared that a more radical design might take longer to build and tune up, although, if we were lucky, it might give greater performance in the end. The design finally agreed on was a synchrotron to produce protons of about three billion electron-volts energy, which came to be called the Cosmotron. It turned out to be quite useful, although it has since been superseded by much larger machines.

By the winter of 1947–48, we had moved into many of the buildings on the camp site and had set up all the facilities and hired the personnel needed to heat, light, and protect a hundred-odd buildings, to landscape and care for a thousand-odd acres of land, and to operate housing for visitors and dining rooms and recreation facilities for everyone. At this point I found my job had drastically changed. Half of my time had to be spent outside the laboratory, increasing the public's knowledge of it. To ensure contact with physicists, I took a more active part in the professional societies. To keep in touch with Washington, I accepted appointments to various advisory panels. I also became a trustee of the newly formed Rand Corporation.

One public task of some interest was to write a portion of the so-called Eberstadt Report on the proposed union of the military services into a single Department of

Defense. The report was requested by James Forrestal, who had been Secretary of the Navy during World War II and had seen the immense strain placed on the war effort by the division of the armed services into two separate chains of command, brought together only at the level of Cabinet and Congressional committees. This division had been feasible in earlier times, when Navy and Army operations were quite separate (although Lincoln at times had had difficulties in getting the two to work together when they were supposed to). In World War II, however, a large number of operations were conducted jointly; in addition, the air arm had come to be deserving of equal status. With his Wall Street background, Forrestal had considerable respect for the abilities of Ferdinand Eberstadt, a private banker of experience and reputation, and asked him to undertake the study of the unification proposal.

I first heard about the study in the fall of 1945, before I went to Brookhaven. Evidently the talks Kimball and I had given about the atomic bomb to Forrestal and the Senate Naval Affairs Committee had made an impression. I was asked to go to Washington to meet with Eberstadt, who wanted to know whether I could make suggestions regarding the interface between the science-technology community and the proposed Department of Defense. I liked him, so I agreed.

By the time the study was organized, I was busy putting Brookhaven together, so my participation was sporadic. I agreed to write a paper on science, the armed services, and government, which could be a supplement to the main report. In my limited time, all I could do was to stress the importance of having a Deputy Secretary of Defense who had scientific training and who would be responsible for military research and development and to urge that an operations research group be a part of the Secretary's staff, with close contact with the Secretary and with his senior military advisers.

At this present distance I wonder whether the unification of the services was not the proximate cause of the increasing militarization of the United States. Instead of enabling the civilian executive to control the military more effectively, it seems to have enabled the military complex to gain in unregulated power.

On the other hand, at that time I had been seeing the wasteful rivalry that flourished when the three services were separated. After VJ day it was first priority with each service to claim a share, preferably a major share, in the nuclear pie. The one application of nuclear science at that time was the bomb. Each service, therefore, wanted its own kind of bomb, with its own special kind of aerial transport to deliver it. The Air Force had an obvious advantage in this respect, but that did not hinder the other services from starting dozens of different projects for different delivery vehicles, all of them expensive, many of them impractical, and each designed to be a monopoly of that service.

In 1947, while I was at Brookhaven, I arranged a visit to Adm. Earle Mills, head of the Bureau of Ships, which was responsible for the design and construction of naval transport. My message to him was that the nucleus could produce motive power as well as bombs and that it would be better strategy for the Navy to produce nuclear-powered submarines than to compete with the Air Force in guided missiles. I do not know whether my conversation had any effect on the Navy's actions; however, a few months later, I met Capt. Hyman G. Rickover (later Admiral Rickover) at Oak Ridge, where he was beginning his long, effective, but stormy career in the development of nuclear submarines. His voice quivered as he outlined to me the tremendous potentialities of nuclear power in underwater vessels.

A chance to assist in the strengthening of civilian control of the military came when, as I mentioned, I was asked to join the board of trustees of the newly formed Rand Corporation. At the end of the war, Ed Bowles, an MIT friend and head of the scientific group advising General Arnold (the group with which Stratton and Shockley had served), had recommended that the Air Force set up a group of scientists to study air-war problems. Bowles believed that the group should be industrially inclined and suggested that it be set up under a contract with some airplane manufacturing company. So Donald Douglas was persuaded to set up a *R*esearch *AN*d *D*evelopment group (thus the acronym RAND) for the Air Force. One of Douglas's senior engineers, Frank Collbohm, who had been with Bowles's group during the war, was chosen to direct the new organization.

It soon became clear that such a body, supposed to be giving unbiased advice to the Air Force, was an embarrassment to the company it was attached to. The competing airframe companies were the first to complain that Douglas Aircraft would have an inside track on contracts, and Donald Douglas began to fear that his company would lose sales if the Air Force had to counter such a suspicion. At any rate, the company soon was trying to rid itself of Rand, which posed a problem.

Neither the Air Force nor the small band of experts

who had already been recruited for Rand wanted the group to have a civil service classification and be moved to some military base. The essential idea of Rand, agreed to by all concerned, was that it should be as free from Air Force pressure as possible, so that it could deliver truly unbiased advice. A better idea was for Rand to become an independent, nonprofit corporation, organizationally separate from both Douglas and the Air Force. Luckily, at about the time this crisis arose, H. Rowan Gaither, a San Francisco lawyer who had run the business administration of the Radiation Laboratory at MIT during the war, had become acquainted with Collbohm's problem. Gaither had been talking with the Ford family about setting up a Ford Foundation and was able to arrange a loan from Ford to support Rand during the transition period while Rand was arranging a contract with the Air Force.

I had heard about Rand from Bowles and others. My first direct contact came when Collbohm asked me in 1948 to be a member of the board of trustees of the newly formed corporation. I accepted, in part because I hoped Rand would have a strong influence on Air Force policy and in part because I was interested to learn whether a detached organization in California could exert as great an influence on the Air Force as the continuation of ORG, by then housed in the Pentagon, had on the Navy. The second question was partly answered when Rand soon found it necessary to set up a liaison office in Washington.

By the beginning of 1948, Brookhaven was well on its way to becoming the unique research adjunct that the cooperating universities had hoped for. None of the major facilities had been completed, but the designs had been ratified and the contracts had been let. We had won most of the battles to make the laboratory a pleasant and stimulating place in which to do research.

My job was changing from that of initiating new policies and setting new precedents to that of running a large, ongoing establishment. During the first year, the laboratory staff had been a relatively small group, in constant contact with one another, stretching to keep ahead of the torrent of things to be done. Now I began to have to influence actions at second or third hand and to hope that what had been agreed on actually got done.

As long as a working group numbers fewer than about a hundred persons, it is possible, although not necessarily easy, to weld its members into a community, in which orders rarely need to be given because each member understands the group's goals and knows his part in achieving them. When the organization is much greater than a hundred people, face-to-face interchange of ideas with everyone becomes impossible, and, unless special effort is made, feedback dries up. The boss doesn't learn how his orders are carried out, and the bossed feels he has no part in the action except to obey. I began to find guards and clerks and shop assistants who had no idea what the laboratory was supposed to do and who therefore felt left out. These discoveries forced me to confront my own problem.

I found I was not fond of being the administrator of a large organization, of being always on display, of having to be careful of every statement lest it be interpreted wrongly—in short, I disliked having to deal with words rather than with facts or actions. And I disliked pushing people around.

Initiation

Perhaps it was fatigue from the two years' effort on top of the wartime stress. I had an indicator of this fatigue when I came back from the winter meeting of the Physical Society at Columbia to discover a slip of paper in my pocket, with a name and address written on it in my own handwriting. I had no memory of how the slip got there or when I had written it. The name was Rosa Streifinger and the address was one in Munich, Germany. Annabelle suggested that Rosa was the name of the maid we had when we were in Munich in 1930, but this suggestion did not dispel the mystery of who in the Physical Society knew our maid of fifteen years before and also had seen her after the war. We wrote to the address and received a prompt reply. It was indeed our Rosa, who had married and had a son. She was as mystified as we were as to how her present name and address had reached us. We still correspond with her and her son Leo. But that whole Physical Society meeting had left no record in my memory.

It took me a few months to reach a decision about my future. Could I go back to MIT without harming Brookhaven's progress and could I do research again? Luckily, by that time, Leland J. Haworth had been persuaded to come to Brookhaven as my deputy. He had been at the MIT Radiation Laboratory during the war and was quite capable of taking over my job. He seemed to be more interested in managing an ongoing organization than I was. In fact he later ran more important things in Washington, such as the National Science Foundation.

That winter I started again to write parts of the text on theoretical physics that Feshbach and I had started ten years before. I discovered that I still could do theoretical physics and still could enjoy the mental struggle of trying to fit concepts together. An added weight in the balance was that my family would be greatly relieved to return to the Boston area. By April I had made up my mind and had arranged to return to MIT by September. It was the right decision for me. I am much more interested in starting things than in running them after they are under way.

Brookhaven continued to prosper and expand under Haworth's leadership. It now is one of the three or four major world centers for nuclear research.

We found a house again in Belmont and moved during the summer. I gloried in my freedom. I wrote furiously at the physics text, I caught up on the work of the Acoustics Laboratory, and I began to get acquainted again with graduate students, who have always offered me the best means of maintaining my mental agility.

But my war-generated responsibilities would not yet let me go. The reorganization of the U.S. military forces into a single Department of Defense had been approved by Congress in approximately the form recommended by the Eberstadt Report, and Forrestal had been appointed as the first Secretary of Defense. An operations research group was in the plan, one of the sort I had proposed in my part of that report. I learned of the plan when Lt. Gen. J. E. Hull dropped in to see me at MIT and asked me to come to Washington to talk to Secretary Forrestal.

General Hull had been head of plans during the war, under Gen. George C. Marshall, and had been in command of the Army part of Task Force 7, which had conducted atomic bomb tests at Eniwetok in 1947. He had agreed to be director of the analysis group to be called the Weapons Systems Evaluation Group (WSEG). The group was to have both military and civilian members and would report to the newly organized Joint Chiefs of Staff as well as to the Secretary of Defense. General Hull wanted me to be deputy director and director of research.

I was somewhat suspicious of a mixed civilian-military group, especially with so high-ranking an officer as overall head. Would the civilians be just window-dressing, with no chance to make independent judgments? Despite these misgivings, the more I talked to Ed Hull, the more I thought the job was worth trying.

Hull had anticipated many of my worries. He emphasized that I would be his deputy, which meant that the military group would take orders from me when Hull was absent and that I would participate in all decisions. He pointed out that his rank would enhance the standing of the group among military people and that he would fight to keep the analysts' freedom of decision.

Hull had already chosen his senior military staff: Vice Admiral Parsons, who had been at Los Alamos; Maj. Gen. James M. Gavin, who had commanded the 82nd Airborne Division in Europe during the war; and Maj. Gen. Earl W. Barnes, who had commanded the Air Force in the Southwest Pacific under MacArthur during the latter half of the war. In addition, there would be six or eight more junior officers, carefully selected for their combat experience and analytic abilities. General Hull wanted me to recruit an equal number of experienced civilian analysts, who would work under my direct supervision.

It was clear that Hull sincerely wanted the group to be a truly integrated one and would do his best to protect it from interservice politics. There was a chance it would work; if so, it would be challenging to be in at the beginning. I went down to see Secretary Forrestal. His face showed many more fatigue lines than when I had seen him last in 1945, but he grew animated as he outlined his ideas about the way the new group would work. He really needed it, he said. So I went back to break the news of my decision to the family, to ask for another leave of absence from MIT, and to start again to look for recruits.

The hunt was easier than it had been during the war. Some of my wartime co-workers had had enough time to become bored with the work they had gone back to. Also, the competition for recruits was less stiff; I could pick up some good men who had been in other wartime operations research groups. But putting the recruit on the payroll was much harder than it had been for our wartime group.

It had been decided that the civilians of the group would be civil servants, which meant that I was handcuffed and leg-chained by civil service rules. I could not bargain about salary, which was predetermined by the recruit's age and education and the number and quality of the civil service positions that had been allotted to the group. Vacations, working hours, and sick leaves were immutably fixed. It took a long time and much paperwork to get a man on the payroll after he did agree to come, and if the recruit did not have a Q clearance, all action ceased while the investigation ground slowly to its completion.

There had been delays and difficulties in recruiting for Brookhaven, but there, my personnel people had been able to figure out a way to get a man to work before full clearance was achieved. At WSEG, everyone was pleasant and a little condescending, but if the job description said

Initiation

that the jobholder should have Q clearance, not a thing could be done until he was cleared. We lost a number of good prospects because they could not wait three or four months. We succeeded in hiring only about five percent of the uncleared men we tried to recruit. We attained a minimal group of analysts that spring by raiding the other military operations research groups, a procedure that hardly generated friendly cooperation.

By March 1949 I was again in the routine of commuting to Boston for weekends. The overnight train service had by then deteriorated from an irritation to a torment, but plane service had improved and some of the younger WSEG air officers were often willing to run me to Hanscom Field in a B-25, noisy but fairly quick. I began to get acquainted with the military members of WSEG and with the postwar Pentagon labyrinth. The senior officers were new to systems analysis, but were interested and hopeful about its possibilities.

Deac Parsons had already been immunized to scientists at Los Alamos. He had not had as much combat experience as had the others, but his contacts with naval planners and with the Military Advisory Board of AEC were solid. We could rely on him to ensure that we had a say on atomic bomb policy. Jim Gavin was less familiar with the Pentagon; he would be valuable for his combat experience as well as for his analytic abilities. His book on airborne warfare had just been published. I suggested that he look into the pros and cons of the tactical use of atomic weapons. I hoped his background and his skeptical attitude would provide a healthy defense against the prevailing faith that the bomb was all we needed to maintain control over the world. Diz Barnes's experience in the South

Pacific would be useful when we came to study strategic bombing, as I was sure we would. It was not so certain that he would stand up for us against the Air Staff, but we would cross that bridge when we came to it.

Ed Hull knew well how to become the father figure. He quickly wiped out class distinctions between military officers and civilians. Senior civilians were meticulously accorded flag-officer treatment, to the confusion of officers at installations we visited. Hull's friendship with Gen. Omar Bradley, then head of the Joint Chiefs of Staff, guaranteed that we would be heard.

Some of my wartime colleagues were persuaded to join us. George Kimball divided his time between Columbia and the group. William Horvath and George Welch came full time and, with me, were the civilian members of the review board, the military members being Parsons, Gavin, and Barnes. This board did the planning for the group and reviewed all group reports. In the meantime, a number of new recruits were filtering through the civil service maze. Before the pressures were to rise, we had a chance to find the best tasks for each member and to instill that modicum of awareness of a common defense against the outside that helps bring group unity.

We would need our unity. The three services were not going to be welded together by Congressional edict alone. Forrestal was wearing himself out trying to get even an outward show of harmony. The youngest service was arrogating to itself the sole control of the bomb and was strongly opposing both the Navy's plans for delivery by carrier-based aircraft and the Army's proposal to develop guided missiles. Every officer in the Pentagon seemed to be pushing his own weapons-development project. The funds demanded for new guided-missile projects exceeded total wartime development costs. Nuclear propulsion, guidance by pigeons pecking at TV screens in the missile, every scheme imaginable and some unimaginable had a proponent. Officers newly come to Washington seemed to prefer scrapping their predecessors' projects and starting pet schemes of their own. When Forrestal finally gave up the task, Compton, who had taken the job of chairman of the Defense Research and Development Board, was unable alone to stem the rush for grants.

The younger officers, during the war, had grown used to having the nation at their disposal; they expected—indeed, demanded—that the support continue. They were abetted by the airplane and weapons manufacturers and by the Congressmen who saw political advantage in whipping up fear of Russia. Their criterion seemed to be what was best for their branch of the service, not what was best for the country. And when Forrestal was replaced by Louis Johnson, they no longer paid much attention to civilian control. It was not until General Marshall took over as Secretary of Defense and the Korean war broke out that a slight degree of order was imposed.

The senior officers of WSEG were generally opposed to this extravagance. But the group was hardly strong enough to take on the rest of the Pentagon by itself. As long as Forrestal was in office, Hull and I could present our case at his morning staff meetings. But Johnson did not have staff meetings. His unspoken order, "Don't call me, I'll call you," brought about the same atrophy of intercommunication that we have seen in more virulent form in higher office in the 1970s.

We had to work through the Joint Chiefs of Staff to do

what we could. We tried to support some of the more realistic missile projects and to discourage some of the crazier ones. Parsons and I tried to persuade the Navy to stop duplicating Army and Air Force missile projects and to concentrate on developments that were uniquely naval, such as a nuclear-powered submarine. We may have had some effect there, but we found that this sort of guerrilla warfare demanded much of our energies. We turned out many brief reports, because that was expected of us. But an operations research group cannot count any piece of work successful unless it influences a decision. While I was with WSEG, we had a few such successes, perhaps as many as could reasonably be expected.

One such involved very little work on our part, and our contribution may have had only an indirect effect on the policy decision, but it is worth mentioning because the decision has been debated since. The occasion was the argument within the AEC about whether or not to develop the hydrogen bomb.

Just as with the original A-bomb, it was not certain whether an H-bomb could be made, whether the nature of the various nuclei involved was such that an effective thermonuclear explosion could be produced. A number of experts, among them Edward Teller, felt strongly that the AEC should go ahead. Many others, among them Robert Oppenheimer, felt that we had gone far enough in developing weapons of mass destruction and that we should not even try to find out if an H-bomb was possible. The question was under heated debate in the Military Advisory Board of the AEC, and we were asked for an opinion. Our answer was to be delivered within a week. The time allowed was not long enough for us to carry out a study,

Initiation

but the decision really depended on gut feeling and general experience, rather than on analyzable data. The officers of WSEG were inclined to vote to go ahead, but they wanted the opinion of the civilians before they made up their minds.

The caucus of the senior civilians was solemn but surprisingly brief. It turned out that we all were agreed that the AEC had to go ahead. A refusal to learn whether an H-bomb was possible was to leave it for someone else to make the investigation. Someone would find the answer; if not this country, then some other. We could also see the inevitable outcome of our recommendation. If the AEC found that an H-bomb was possible, the pressure would be irresistible to build and test one—and none of us liked that. Nevertheless, we felt that in the long run knowledge is safer than ignorance, particularly self-imposed ignorance.

One never can be sure how much influence one group's opinion has on a decision as weighty as this one. However, the fact that our opinion was unanimous may have had some effect on the decision to go ahead.

Another action that had some effect involved a great deal of work. The Strategic Air Command had been set up to deliver A-bombs, to be what has since been called our deterrent force. At that time, their planes were still propeller-driven bombers. SAC wanted a new bomber that would fly higher, faster, and farther. What was needed was a long-range jet bomber; only the medium-range B-47 was then operational. But the only long-range bomber that could then be produced was the ungainly B-36, with six propellers and the possibility that two jet pods could be attached later.

It was a huge plane requiring a crew of more than a

dozen men. It was one of the first to be pressurized and could thus fly higher than anything else we had. The Air Force was urging the purchase of enough B-36's to fit out SAC at a cost that seemed, at the time, astronomical, particularly as these planes were to be scrapped as soon as a true jet bomber was built. Neither the Army nor the Navy was happy about the Air Force's preempting such a large part of the military budget, but the airplane manufacturers were all for it and Secretary Johnson was on their side.

The controversy had reached the floors of Congress, where the proponents of the Navy claimed that the B-36 was vulnerable to fighter planes and would never reach its targets, while the proponents of Air Force plans proclaimed the B-36 as a precise and reliable weapons system. Because of the urgency and the need for a high-level resolution of the question, WSEG agreed to undertake a study of the matter and to report within six months. I appointed Bill Horvath as leader of the project, which eventually came to occupy the full-time efforts of the entire group, as well as those of many outside groups, including the Air Force and the Navy Operations Research groups, a part of the Aberdeen Proving Ground, and some of the staff of Rand.

We had hoped to break our group in on some less controversial study, but this problem was too urgent to evade. The task was much more difficult than any the Navy group had attempted during the war, when we knew at almost every moment the enemy's capabilities as well as our own. Here we had to guess how well our crews could handle an untried plane and to estimate as well as we could what defenses a possible enemy (the Russians, for example) might have a year or so in the future. We had to

Initiation

extrapolate from World War II data on the likelihood of a bomber's surviving antiaircraft and fighter defenses to a situation in which the bombers were flying higher and faster. We then had to estimate what fraction of the bombs would land on target. The changes from World War II conditions were not yet as great as they have since become, so our job was not as much sheer guesswork as it would be now. Nevertheless, all we could do was to produce a range of answers, from optimistic to pessimistic, and to give the results as comparisons with estimates of the performance of the bombers then in use, under like conditions. This procedure, now standard practice, is known now as sensitivity analysis.

The biggest uncertainties came in the estimates of Russian defense capabilities, which we obtained from the Central Intelligence Agency. Our Navy group had used data from Naval Intelligence during the war, but this had been fairly hard data, obtained from breaking the German radio code or from questioning prisoners. Now we had to use figures obtained from much less dependable sources: spies, chance observations of Russian equipment, comments from occasional defectors, or obscure paragraphs in Russian papers. All these fragments were put together by people with whom we had no direct contact and were "filtered" by still others, using procedures carefully concealed from us. The more I saw of the estimates, the more I wondered whether there was any quantitative system in the "filtering" or whether the reports were the consequence of the political predilections or the daily state of mind of the filterers.

I suppose it is not safe to allow all users of intelligence data, even when sworn to secrecy, to see how the data are gathered and analyzed, so they can judge the accuracy of the information for themselves. But the reports we did get did not produce confidence. Estimates of numbers and of effectiveness would vary widely from month to month, sometimes by factors greater than two. We suspected that estimates were usually put on the high side, just to be safe, but were occasionally reduced in a sudden urge to be realistic. All we could do was to compare the estimated performance of the B-36 against that of the bomber then in use, for both the high and the low intelligence estimates of enemy capabilities, and hope that the spread was not so great as to make the results meaningless.

Political pressures, we found, were likely to be even more serious than the vagueness of the data. Our first runs, using an estimate of enemy capacity midway between the intelligence extremes, indicated that the B-36 was not appreciably better than the bomber then in use and that under some circumstances it might even be less effective. The major detrimental factor was the large size of the new airplane; it was considerably easier to hit, and its increased speed and altitude advantage over the conventional B-50 did not seem to be enough greater to neutralize this disadvantage. We knew that if we came out with this sort of result, without preparing the way, we would simply unite the Air Force against us and cut off future cooperation on other studies. We would have to get at least a few influential fliers to agree with the results before the report was published.

The best sort of report recommending a policy decision is one that is history by the time it is published, agreement having already been achieved. Turning out a report without close cooperation with the group affected

Initiation

by the report is usually an exercise in futility. We therefore had to get in touch with the Strategic Air Command and, if possible, persuade them to participate in the analysis. The conclusions would then be available for them to act on before the report's publication. Indeed, they could then say that the conclusions were their idea in the first place. Our group was not looking for credit for its work; we just wanted action. So Ed Hull and I went out to Omaha to confront Gen. Curtis LeMay, then SAC commander.

It was an interesting session. LeMay sat across a big table from the two of us, flanked by his staff. LeMay's argument, which he presented clearly and at some length, bypassed the question of the greater effectiveness of the B-36. He brought in a factor that tends to be forgotten because it cannot be quantified, the problem of morale. He emphasized the psychological difficulties of those who guard, who spend years in apparently useless watching and then are expected instantaneously to do the right thing when an attack occurs. The best way to counter the lethargy of inaction, he suggested, is to give the guards a sequence of new things to play with to keep them on their toes. Of course, he made these points in more seemly and prolix terms.

The problem LeMay posed was a real one, whether or not his proposed solution would solve it. Having studied reports of the Pearl Harbor disaster, as well as many other historic examples of initial blunders from Trasimene to Austerlitz, I believed and still believe that in first clashes nothing will go right for either side and that the best defense is an ability to learn from experience. There was no need to try to debate LeMay's argument. I simply said I could see that the general would make a good systems analyst and that I was sorry we couldn't recruit him to work on the B-36 study. I was rewarded with a trace of smile on LeMay's face and an approving murmur from his staff. I went on to say that we needed assistance in making the study a realistic one and that I hoped he could assign two or three of his staff to come to Washington to help us.

This was not selling out to the opposition. I was sure his men could not alter the nature of our conclusions; I wanted them there so we could learn in advance the arguments that might be used against us and so we could word the report to avoid misunderstandings later. And if the Strategic Air Command should act on the results before the report was published, so much the better.

The session continued in a friendlier mood. There was a SAC promise to consider assigning some staff members to our group for a few months and a suggestion that perhaps our people would like to fly in some of the prototype B-36's the Command was then testing. This was just what Hull and I had hoped would be offered. By the end of the day, LeMay and his staff seemed to be convinced that we were not adversaries, but were really trying to determine the facts.

A few weeks later, most of our group went out to Fort Worth, where the prototype B-36's were based. Three of them were sent up, each with a third of our contingent aboard. The planes were to fly down to southern Florida, make a simulated bombing run on Birmingham, and then return to Fort Worth. That day also was an interesting one. We all had to put on parachute packs and receive instructions in their care and operation. Pressurized cabins were a new thing then, so we were warned to stay away from exit doors and were told a cautionary tale about a door that had blown out the previous month, taking a flight sergeant with it. We all had a chance to look at a pile of aluminum at the side of a runway, the remains of a B-36 that had crashed on takeoff the previous week. By the time we took off, we were suitably impressed.

We spent most of the first hour of the flight in the bombardier's bay, enjoying the view. This was the first time I had been above 40,000 feet. The air was clear, and we could see from Texas to Alabama. But then the report came that the radar dish wasn't rotating and, a little later, that one of the six engines was acting up. We went back to the radar compartment to learn more about the problem. The dish rotor was evidently frozen, so we could not make the simulated bombing run. As for the ailing engine, the controls were new and complicated, so the crew decided to shut down the engine rather than to experiment with it. I learned that each engine had a control device with four vacuum tubes; if any one of the tubes broke down or burnt out, the engine was out of commission. I began to feel that the B-36 was just too full of new gadgets to be a dependable piece of military equipment.

There was a bit of a flurry just before we touched down at Fort Worth, and we fishtailed erratically down the runway rather than coming smoothly to a stop. We then learned that a second engine had failed just before we landed. Therefore the usual procedure of reversing the propellers had given us a bad list to port that had to be counteracted by spasmodic brake and rudder action. Altogether, it was not the most convincing demonstration.

Back in Washington, we labored for months, piling up working papers, computations, and simulated runs against various kinds of defenses. The final results bore out our first runs and won the grudging acquiescence of the officers lent us by LeMay. The B-36 promised a somewhat better performance then the existing bombers against a few defenses, but the average behavior was not enough better to justify the cost; the plane's sheer size and complexity were a real handicap. It would be better, we felt, to wait for an all-jet bomber. A few B-36's should perhaps be purchased to keep up the morale LeMay had stressed, but a large purchase would be unwise.

By the end of the study, we felt we had converted enough Air Force officers to make our recommendation stick; but we still had to convince the politicians, particularly Louis Johnson. General Bradley, as head of the Joint Chiefs of Staff, had mentioned the study to President Truman. So a request came through from the White House asking us to set a date, preferably soon, for briefing the Cabinet.

At the session with President Truman and his Cabinet, Hull was to read the report, but Horvath and I were to be present in case there were technical questions. We had worked hard to make this version of our report understandable and brief, without omitting essential steps in the logic and without oversimplifying the conclusions. Most of the Cabinet were there, but I remember only Truman, Secretary of State Dean Acheson, Secretary of Defense Johnson, and Secretary of Labor Maurice Tobin. I remember Tobin only because he fell asleep as soon as Hull began to read. Truman and Acheson listened carefully, and Johnson stayed awake but seemed more interested in watching faces than in listening. When Hull had finished, Acheson asked a perceptive question; then Johnson turned to Truman, beamed and said, "There, I told you they'd say the B-36 is a good plane." Truman looked disgusted and snapped, "No, dammit, they said just the opposite." So at least two of our audience got the point.

Again, it is difficult to say what influence our work had on the ultimate decision. Certainly Air Force enthusiasm for the plane diminished while we, with some help from SAC, were making the study. LeMay later wrote Hull praising Horvath and Welch for this and other studies that led to an improvement in SAC. More B-36's were purchased, but they were not brought in the large numbers originally requested. By the time our final report was presented to the Joint Chiefs, we felt it had already affected policy, and we turned with relief to other problems.

By that time, too, I felt I had done all I had been expected to do for WSEG, and I set about returning to MIT. I had originally told General Hull that I would get the group started, which was what I seemed to be able to do best, and that I thought I could do that in about a year and a half. I felt, and Hull reluctantly agreed, that I could with good conscience go back to MIT by September 1950, in time for the fall semester. Between us we persuaded my good friend Bob Robertson to take over as director of research, and the group went on to greater things, particularly during the Korean war. I spent a few days a month with them, as a consultant, until Robertson and then Hull left, and the group began to change.

In fact, all the operations research groups working for the armed forces in Washington were changing, as the experience of World War II receded. The problems the groups were asked to work on became more tinged with interservice politics, and the data underlying their evaluations became ever more vague. Our intelligence agency might be able to count the number of Russian A-bombs and missile silos, but neither we nor the Russians really knew what these weapons could do. Calculations of nuclear warfare capabilities became less and less scientific and more and more astrological.

As for conventional warfare, we seemed to have taken a wrong turn. We seemed to have allowed the Air Force to persuade us that we could win wars without any of our citizens being killed, by substituting gadgets for soldiers. I am not sure how many more Vietnams we will have to fight before we are persuaded that we should fight no war unless we feel so strongly about it that we are willing to give up lives—our own lives—to win it. World War II was the last time we felt that way.

As time went on, each of the service evaluation groups was caught between the military preference for gadgets rather than plans and the growing lack of quantitative data on weapon effectiveness, either ours or a potential enemy's. The three armed services became too occupied in competition among themselves to appreciate objective advice. WSEG's downfall was roundabout. Bill Shockley, who had become director of research after Robertson left, decided that his civilian staff could no longer operate under civil service. A new, not-for-profit corporation, the Institute for Defense Analyses (IDA) was set up to administer the group, as Columbia had administered the Navy group during the war. The two moieties of WSEG began to act independently, and eventually the civilian section had to move out of the Pentagon. By the time Robert McNamara arrived as Secretary of Defense in 1961, WSEG had sunk so far down the chain of command that he preferred to start his own analysis group, recruited mostly from Rand personnel.

If ever another major war breaks out and the armed services again realize that they need analytic help, they might best start afresh, with newly organized groups, rather than trying to reinvigorate the members of the existing groups in Washington. This is not to say that these existing groups have not made valuable contributions in their time; it just appears that advisory groups such as WSEG and OEG have a finite useful life. NDRC superseded existing groups and laboratories at the beginning of World War II, thus bringing in innovators and analysts who would not ever have been willing to bury themselves in the earlier groups or laboratories, and this is a useful precedent.

When I went back to MIT, I also returned to the board of trustees of the Rand Corporation, from which I had taken a leave of absence while I was with WSEG. I believed that Rand had a chance to keep the Air Force within bounds in its weapons development. It was located far from the Pentagon, had freedom to choose its own investigations, and was capable of looking into problems of technology and economics as well as those of tactics and individual weapon evaluation. For the rest, I decided to devote the time I could spend on operations research to developing its applications to peacetime operations, such as to industrial and urban problems. For the time being, however, I wanted to get back to teaching and research.

9
Instigation

Back again at MIT in the fall of 1950, I wanted to stay this time. I had had my fill of the Pentagon, although I was still willing to spend a few days a month there consulting. Greater immersion in the Washington atmosphere, then one of Red-baiting, would end either by repressing all my initiative or by branding me as a fellow-traveler, in either case destroying my usefulness. I felt I could be more effective by operating from a base at the Institute. There I could do what I wanted to do, rather than what someone else thought I should do. I still hankered to turn out some research of my own. I realized, of course, that my chance of making some notable discovery was by then small; nevertheless, I still wanted occasionally to pit my mental muscles against some puzzle, to shove formulas around until they fitted together into an understandable picture of some small part of nature.

There were also more important things that I felt able to accomplish and that could best be done from MIT. It was clear that science and technology had reached a new level of importance in the world and that new scientific concepts and the new devices they generated must play an ever larger part in people's lives. Nuclear physics had recast military strategy; it might also revolutionize our energy economy. Electronics was just beginning to show what it could do in spreading ideas and emotions and in helping us to think faster and more accurately. Every such technical advance brings social changes, and every social stress can be either aggravated or alleviated by the way in which the advances are introduced. Convincing examples had occurred during my own youth, with automobiles and radio, for example. But the pace was faster now.

Many more people trained in science and scientific thinking would be needed to discover new concepts and create new devices, but also to help control the way they were introduced. I wanted to have a hand in training these people, in finding those with the judgment and initiative to become leaders, in teaching and motivating them, and in seeing that they found jobs in which their capabilities would be properly used. There would be need for research physicists, and I could play my accustomed part there. But a more general kind of scientist was going to be needed, one with thorough training in the scientific method but with an interest in the interactions of various new technical developments as well as in the conflicts between these developments and the social system. Scientists would be needed who could use experience gained in some specialized research to reach an understanding of these broader interactions. James R. Killian, who had by then succeeded Compton as president of MIT, heartily supported these ideas.

My work in operations research showed me how dangerous a team composed entirely of narrow specialists could be and how necessary it was to have a mixed team, accustomed to exchanging ideas across disciplines, in order to introduce successfully some new instrumentality. Education of these specialists-with-generalist-predilections would necessitate weakening the barriers between the departments at MIT. It would mean that the civil engineering department, for example, would have to tolerate and value a colleague who occasionally worked in electronics; that the physics department would have to be willing to let one of its graduate students write a thesis under the supervision of someone in the economics department. To bring about any such relaxation of provincialism would first require sales effort and then time and energy to create new organizations. I was experienced in that sort of work, I believed.

But first I had to fit myself back into the physics department. The stint with WSEG had not required me to drop out completely. Teaching a course had been impossible, but I had been able to advise graduate students and to supervise the thesis research of a few of them. A little time had been squeezed out to work with Feshbach, planning the scope of the ever-growing treatise on theoretical physics. Work in acoustics had already begun; Bolt and I had written a long article on room acoustics, and the revision of *Vibration and Sound* had been completed on weekends and in the bedrooms of the Federal Express, to and from Washington. During my Brookhaven days, the Acoustical Society had elected me its vice-president and then its president-elect. I became its president at the Society's meeting in Cambridge in November 1950. The American Physical Society elected me member of its council, which helped me to become reacquainted with physics colleagues in other universities.

In the fall of 1950 my first task was to get back in touch with physics students, particularly with the graduate students. There were now many more of them at MIT, and they differed from their forerunners in the thirties in ways both obvious and subtle. Because of the enormous expansion of the funds to support scientific research and education, and in particular because of the existence of the GI-bill educational grants, more young people were emboldened to apply for graduate work in physics at MIT than we had room for. Before World War II, our choice had had to be limited to those who could afford to come. Now, in principle, we could admit anyone who appeared capable of passing our courses and achieving an advanced degree, without considering whether his or her parents could pay the cost. As a result, the student body had an increasing number of so-called minority representatives: blacks, women, citizens of other countries, and so on.

This separation of admissibility from affluence posed a basic policy question: should the physics department continue to expand indefinitely, admitting everyone we thought could pass our courses, or should we apply some limit to the number we admitted? My colleagues in the department felt, and I agreed, that we should not dilute our effort, that we should concentrate on educating an elite, thoroughly grounded in all branches of physics and capable of carrying out research of high quality. Other universities could adopt production-line methods; we wanted to continue turning out a few of the best, although the few turned out to be three times as many as we had coped with in the thirties.

Deciding who was to be admitted was therefore difficult. Between four hundred and five hundred applications were received each year for graduate work in physics. Only about one in eight could be admitted, if our estimate of fifty students a year as the optimal number was to be held to. Few but the best from undergraduate schools applied to MIT, so we were picking the best from the best, a hard task. This task was further complicated by the difficulty of appraising many minority students and those from abroad. If we also wanted to educate the best of these groups as well, we would have to weigh the student's ambition against his lack of preparation or his difficulty with the language. In some cases, we had to rely on hunches and hope that later we would not have to crush the student's spirit by flunking him out. We took some of these gambles each year.

One student we gambled on came from Howard University, the first who had applied from that black college. When he arrived in September 1949, the problem was to persuade him to take mostly undergraduate courses for his first year, without discouraging him unduly. Luckily, he was interested in acoustics. The other students in the Acoustics Laboratory were helpful and not condescending, so he managed to scrape through his first year. It took him five years to get his Ph.D. He stayed on for two years more, working in the Acoustics Laboratory and then went on to Los Alamos with our strong recommendations. He did well there and is now back at MIT as a full professor of physics.

Still another student we took a chance on came from Turkey. Huseyin Yilmaz had been a shepherd boy who battled his family to let him go to school. He eventually earned a Turkish government scholarship to the National University at Ankara. I heard of him from a friend in the American Embassy in Ankara. My friend wrote that the young man seemed to have promise and that money could be found to fly him to Boston if MIT could pay him enough to stay alive while he studied physics. We decided to take the gamble and accepted him. He had trouble with his visa and did not arrive in September. One Sunday morning in October he rang the front doorbell of our home in Winchester, a few miles out of Boston. He had an Istanbul newspaper under one arm, a box of Turkish delight under the other, and stars in his eyes. He had just landed that morning, had found his way to MIT and thence to our home, an achievement requiring intelligence and initiative of the first order, as anyone familiar with Boston public transport will admit. He obtained his Ph.D. degree in just three years, passing all his courses with top grades. He is now president of his own small electronics company just outside Boston.

There were other differences between the physics students of the fifties and those of the thirties. In the thirties, a faculty could be sure that anyone who went into graduate work in physics liked physics more than money or renown. But by the fifties, physics had become the glamour field. Among those choosing physics in the fifties were some looking for fame and even power, in addition to the joys of pure research. A little of this was probably present in the minds of most physics students; it changed their attitudes and behavior in various ways. In some students, this attitude was manifested in a tendency to stay away from wide-ranging curiosity and in a preference for safer specialization, particularly in a popular subfield, such as nuclear physics. It increased the atmosphere of competition to some extent; students were not as mutually helpful as they had been in the thirties.

Another, deeper difference disturbed me more and more as the fifties wore on. The surface symptoms were that the students were more polite to their professors; they seldom disagreed with us, and they usually tucked a "sir" somewhere inside each sentence. A teacher usually had to work hard to get an argument started in class. Some of these symptoms may have been linked to the influx of ambition and conventionality that accompanied the rise in status of physics. But as time went on, I became convinced that it was caused more by the ubiquitous military influence on our country's youth.

In the fifties, a large number of the graduate students had been in World War II or had been drafted soon after. They had gone into the Army at the bottom, where everything was decided for them and where disagreement was strictly forbidden. I could see the damage, at first vaguely and then, as time went on, more and more clearly. The draft affected students in many ways, most of them detrimental, the more so the more brilliant the student. Those who went into the Army spent a third of their most creative years in an environment designed to remove curiosity and initiative. Most breakthroughs in physics are made by scientists before they reach their early thirties, remember.

Those students who relied on student deferments to keep out of the Army were locked into school. Perhaps not all of them would have benefited by taking a year off before college or graduate school. Nevertheless, the fact that none of them could take leave from formal education without being drafted lay in the back of their minds, an indistinct feeling that they were deemed useful only in war. In the fifties, all that this situation seemed to produce was an added conformity and a use of politeness to hide a feeling of separation. By the sixties it began to show as bitterness, and the lack of accord became outspoken.

But reestablishing my connections with colleagues and students was only part of my general plan for the next decade. I wanted to finish the book on theoretical physics and to generate whatever research ideas I could. Since 1947 Feshbach and I had been writing parts of the book and discussing its scope. Now was the time to delimit it and fix its pattern. The plan was broad but simple: we wished to include enough to enable the advanced student to work in any of the areas of theoretical physics that had arisen since 1925.

Theoretical physics has several ways of describing and predicting the behavior of matter. One way is called classical dynamics. It deals with the motions of particles and discrete bodies. Another is the theory of fields and waves, which deals with forces and other effects that pervade the space between the discrete bodies of classical dynamics. The century after Newton concentrated mainly on developing dynamics, but since then field theory has forged ahead, accelerated by the growing realization that the mathematics describing one sort of field is very like that for another. The equations for gravitational potential turn out to be the same as those for fluid flow and for electrostatic fields; acoustic waves, optical waves, and indeed all electromagnetic waves obey the same wave equation.

This part of physics had been known before 1925, and many excellent texts gave details. Then quantum theory disclosed that for particles of atomic and nuclear size the appropriate equations are not those of classical dynamics, but are more closely related to the long-known wave equation, a field equation. Thus field theory became the most important division of theoretical physics.

So our treatise was to be about field theory—how to devise and solve the equations for the various acoustic, gravitational, electromagnetic, thermal, or quantum fields. We wanted to put in logical order and readable form the manifold techniques of calculation of the wide variety of solutions of the remarkably small number of equations that served to describe all these fields. We first had to show the connection between these equations and the multitude of physical phenomena they could be used to describe. The book was to be a physics, not a mathematics, treatise, and we wanted to link the equations with the physics throughout. Then we had to discuss the various guises in which the equations could appear—in a differential or integral form or as a statement that something had to be maximum or minimum, a variational principle. The next step was to explain their equivalence—why, for example, minimizing the field's total energy can be equivalent to a differential equation governing the field's behavior at every point.

Next we felt we should outline the mathematical nature of the analytic functions, in terms of which the solutions are expressed, and the intricate ways by which these functions can be made to correspond to some actual situation. Then each of the differential equations for the fields was to be discussed in detail, relating each term to the corresponding property of the field it governed, so the student could see why such a small number of equations could represent the behavior of so many kinds of phenomena. Solutions, to be useful, have to be expressed in terms of appropriate coordinates, but the equations take on different forms in different coordinates, so the underlying unity of these forms had to be demonstrated. After those preliminaries, we planned to show how the boundary conditions—the presence of electric charge or the influx of heat, for example—at the surface of a region affected the form of a field within the region and how those effects could be calculated.

We wanted to make all this material as readable as possible, at least for a person fluent in the language of mathematics. We wished to explain in advance the reasons for each step in the exposition, rather than simply to present a bald and unconvincing sequence of mathematical theorems, with the connective tissue of motivation and relevance left to the reader to puzzle out. I even worked out a way to draw surfaces, illustrating the equations, so they could be viewed stereoscopically, as if in three dimensions. Above all, we hoped to give the student a glimpse of the intricate, polyphonic beauty of field theory.

By the time we had proceeded this far, we realized we already had enough material to fill a book of a thousand pages, and we were only halfway through our plan. The detailed exposition of the various kinds of solutions of the several differential equations, for many boundary conditions, turned out to fill another volume. By the end of 1952 we had finished. We carried the manuscript to the publisher in New York in two large suitcases, where it was received with awe and dismay. Proofreading all the tables and equations was more toilsome than the writing, but by the fall of 1953 the two-volume treatise, called *Methods of Theoretical Physics,* was out. Income from its first year's sales topped that of any other work in the publisher's college list—but, of course, few texts sold for as much as $25 a volume then. A Russian translation was published; but we received no royalties from that large distribution. Now, twenty years later, the royalties still pay for the family's vacation trips. The book may also have calmed the doubts of some *illustrissimi* as to whether I still had standing as a physicist, for I was elected a member of the National Academy of Sciences in the spring of 1955.

Feshbach and I took turns giving the lectures in advanced theoretical physics. While the text was being written, we could try out the clarity of our exposition; later we had help from the students in locating errors in the text. I also took my turn at giving some of the other theoretical courses, such as electromagnetic theory and, occasionally, quantum theory or relativity. Teaching still was a pleasure. To break through the student's apathy and/or politeness, to stimulate argument and through it understanding, requires preparation and some histrionic effort and is rewarded by an occasional grin or a gleam in an eye. When bright students come around after class with tough questions, I feel I am getting them to think for themselves, and that for me is the payoff.

The final stage of graduate teaching is the supervision of thesis research, where education reaches the person-to-person ideal of Mark Hopkins. Here the instructor must guide and suggest but not lead. The student now is making science, not just hearing about it. It is this research experience that he can carry over from one specialty to another. That he has, himself, achieved new understanding in physics makes it easier for him to do the same in some other science if he chooses.

Instigation

I had been supervising one thesis on weekends while I worked at WSEG. The student was a Chinese girl, quiet, industrious, and capable. Her subject was the behavior of electrons in a highly compressed gas, an extension of my prewar research on matter in the interior of stars. She finished the work at the end of my first year back at MIT. She is now teaching physics at the University of Shanghai.

During that first year back I agreed to take on the supervision of seven more doctoral theses. One of them was an extension of the theory of the transmission of sound in ducts, which I had worked on before the war, and an experimental verification of the theoretical results, carried out successfully by James Young, the first student from Howard. Two others were calculations of the scattering of electrons and atoms, applying some of the variational methods of computing fields that Feshbach and I had been writing for the treatise. The other four used electronic computers, one to calculate the behavior of cosmic rays in the earth's atmosphere, another to compute atomic wave functions, and a third to predict the flow of water from one dam to another along a river. The fourth thesis was on the sources of errors in analogue computers.

These four theses were a reflection of my growing interest in electronic computers, part of my bent toward instruments and research that could be applied to cross-disciplinary problems. By 1950, the prototype of our present high-speed data processors were just beginning to be fitted together. I could appreciate the excitement that would be involved in learning how they worked and could anticipate some of the ways in which they could ease and speed the work of verifying and then extending any quantitative theory. Much of my own research time had been spent in calculating specific solutions of general equations, first to see whether the equations I had chosen did correspond to reality and then to work out what other phenomena they would predict. Whatever else the new computers might do, they could enormously extend and speed these sorts of calculations.

I had always been interested in the production of tables of values of solutions of the more usual equations—the sort we dealt with in our treatise—so they would not have to be worked out anew each time they were needed. In the thirties I had been glad to encourage Arnold Lowan to produce more tables of solutions of field equations. Experience with the Bush differential analyzer and with the IBM card sorters during the war had given me hints of the potentialities of automated computation.

While at Brookhaven and later, with WSEG, I began to hear of developments in digital computation, of an all-electronic computer called ENIAC, and of some brilliantly simple ideas Johnny von Neumann had about the way computers should be organized. As usual with Johnny's ideas, as soon as he had voiced them everyone began to wonder why someone had not thought of them before. He pointed out that, aside from communication with the outside world, the internal workings of the computer had to be of three kinds. There had to be a memory, where the numbers to be used in the calculations, the intermediate steps, and the final results were stored until needed. There also had to be a processor, where the appropriate numbers from the memory are added, subtracted, or otherwise processed. And, of course, there had to be a controller, with a sequence of instructions, called the program, to be

Instigation

fed sequentially to the processor so that the desired calculation is carried out.

Von Neumann's brilliant suggestion was that the program could be stored in the memory, along with the numbers to be processed. Furthermore, he pointed out that these instructions, if they were in memory, could be modified by the processor, or changed in the light of what had happened to the calculation up to that point. For example, if the intermediate result were larger than a given amount, one set of further instructions would be followed; if smaller, another set would be in control. Thus the computer could decide what to do next, on the basis of what it had just done, rather than waiting for the much slower human operator to decide.

These two simple ideas—arranging for the processor to manipulate instructions as well as add and subtract numbers and arranging for alternative instructions to be chosen by the computer on the basis of the current status—have made possible the immensely flexible and useful "electronic brains" of today and the still greater possibilities of tomorrow. They also have engendered an entirely new and influential kind of intellectual worker, the computer programmer. The processor is a very simple-minded piece of electronic gadgetry. It can do very few things. It can copy out a number residing at a specified point in memory; it can add it, subtract it, or compare it to another number and then return the result to a specified point in memory; and it can do these few things very rapidly and with superhuman accuracy.

Thus the instructions telling the processor what to do next must be both quite simple and immensely detailed, and, since they are to be placed in memory, they must be numbers. Instructions in the code the processor understands are said to be in machine language. Programs in machine language must specify each step the processor must carry out, exactly where each number is to be found in the memory, where the result is to be deposited, and so on. Programs in machine language, for even simple calculations, have hundreds of instructions. If any one of them is incorrect or in the wrong order or contradictory, then the programmer has to devote hours or days to finding the wrong instruction and correcting it, a process known as debugging. With the early machines, it usually took a week or more to write out and debug a program for even a simple job.

By the time I returned to MIT in 1950, one of these new instruments, called Whirlwind, was being completed by a small group of enthusiasts under Jay Forrester, an intense and persuasive young engineer. The money for the project was supplied by the Office of Naval Research. Whirlwind was not the first of the stored-program machines, but it was the first to have a magnetic-core memory, an invention of Forrester's that turned out to be so rapid and dependable that it is now standard on nearly all electronic computers. Each tiny ring core, constituting one "bit" of memory, must be threaded with three wires, interconnecting it with others in a woven mesh of cores and wires. Since a modern memory contains hundreds of thousands of bits, the painstaking job of weaving requires a great amount of delicate hand work. The demand for core memories has built profitable factories in Hong Kong and Taiwan, where labor is cheap.

Forrester's group was so busy building and improving Whirlwind that it had little time to explain its workings to anyone else. Of course, if any faculty member should become interested enough to take the time to learn programming, there was available time on the machine that he would be welcome to use. But Forrester's group had scant time itself to arouse interest among the faculty or students or to teach them once their interest was aroused. Furthermore, the Navy contract did not include funds for this sort of missionary activity.

I do not now remember which of us took the initiative, but my appointment calendar records that Forrester and I had lunch together within a month after I had returned to MIT. Because he was so busy developing Whirlwind, he was quite willing to have someone else take charge of the missionary work. It was a job I was willing and able to do. It had to be taken in steps. First, I had to find a group of faculty members who were interested in exploring the applications of the computer in the wide variety of the disciplines represented at MIT. But this was not enough, for the faculty members, even though interested, would not usually have time to do their own programming. Therefore I had also to interest graduate students who did have time to do the programming and who would work under the direction of the faculty members. In the end, both students and faculty would learn how to use the computer.

The situation in the field of computer studies was a little like that of quantum theory in the twenties, when younger people were able to learn the subject without having first to unlearn the outdated concepts that handicapped the older generation. Facility in programming, as in any other mathematical technique, is a bit like facility in riding a bicycle or skating; it is easier to learn in youth.

But we could not attract students to learn to work with computers unless we had funds for assistantships, so that we could compete with other sources of student support. The competition was with the established departments, with their departmental fellowships and assistantships, and not with the Institute's administration. I knew President Killian from the thirties, when he had been editor of the *Technology Review,* and I had written popular articles on the new physics for him. And Killian's chief assistant was the provost, Julius Stratton, previously a member of the physics department, with whom I had bicycled through England in 1930 and who had the office next to mine during the thirties. If I had a sensible plan and was able to raise the money to implement it, Killian and Stratton would be willing to let me try to nibble away at departmental walls.

I found a number of colleagues from different departments who were interested and willing to help. With Stratton's approval, by December I had become chairman of the Committee on Machine Methods of Computation, composed of nine faculty members, including Forrester. We established a weekly seminar on computers, which attracted students and more faculty members. Next I went to my friends in the Office of Naval Research and persuaded them that they should supplement the funds they were giving Forrester's group to build and improve Whirlwind with a relatively small additional amount to pay for some research assistants to explore the possible applications of Whirlwind to various scientific fields. By the spring of 1951 we had the grant.

The committee was able to select a half-dozen students from as many different departments for our first group of research assistants who would start work the following fall. They and the students recruited in the next two or three years had an exciting time searching out new ways to use computers. One of them went on to Los Alamos to use the machines there in quantum mechanical calculations; another is now a professor in MIT's Sloan School of Management. Some stayed on to work in what became known as computer science; of these, one is now associate head of the Institute's electrical engineering and computer science department. Within five years, the demand for computing services at the Institute had outstripped the capacity of Whirlwind.

Meanwhile I had made contact with others in the thin but growing ranks of computer buffs. People interested in computers were of two sharply differing types: the engineer type included those wrapped up in improving the speed and dependability of the electronic circuits and other equipment (the hardware), and the mathematician type included all those who concentrated on the logic and the human compatibility of computer programs (the software). The separation between the two types of people widened with time, both in location and in interests. By the end of the decade, most of the hardware specialists were working for computer manufacturers, whereas most of the software people were still in the universities.

The dichotomy was apparent in a committee I joined in 1951, the Applied Mathematics Advisory Committee (AMAC) of the National Bureau of Standards. It had originally been set up to counsel the head of the Bureau's Division of Applied Mathematics, which had recently been put in charge of developing machine computation, in addition to its original task of preparing tables of mathematical functions. It was this organization that had supported and supervised Arnold Lowan's tables project in the thirties. AMAC was about equally divided between hardware and software proponents. Eventually the division split in two. One part built and operated the two Bureau computers, one in Washington (SEAC) and one in Los Angeles (SWAC). The other used the computers to calculate tables and to develop programs.

My appointment to AMAC came about because Ed Condon, under whom I had worked at Princeton, had become director of the Bureau of Standards in 1945. I had seen little of him since my days at Princeton, until my work at WSEG in Washington gave me a chance to visit him in the director's house on the old Bureau campus off Connecticut Avenue. He had done a lot of things since 1930. He had directed research for Westinghouse, worked with the Manhattan District for a while during the war, worked with Szilard to put atomic energy under civilian control, participated in the bomb tests at Bikini, and then, at the urging of Sen. Brian MacMahon, taken the job of directing and expanding the Bureau. He was just the same as he had been twenty years earlier, rough-spoken, friendly, outgoing. He was full of plans for making the Bureau what it should be, the science arm of the federal government, and he was loudly critical of governmental red tape. Emilie, his wife, was active in the causes that were coming to be denounced by the Red-baiters. His sons were beginning to follow in their father's scientific footsteps.

In the fifties, Ed was beginning to be the target for a barrage of innuendo and accusation from Senator Joe

McCarthy and his henchman, Rep. Richard Nixon. He was fighting back but was getting discouraged, fearing that, if he stayed on, the Bureau would suffer and that many of his plans for enlarging the scope of the Bureau would never come to fruition. He had already set up a branch of the Bureau at UCLA, where the SWAC was housed, and was talking about moving the main body of the Bureau from Connecticut Avenue out to Gaithersburg, Maryland, where there would be more room. He was excited about the potentialities of electronic computers and, when he found I also was interested, he asked me to become a member of AMAC. Before I had a chance to attend a meeting, he had left the Bureau under pressure from Congress and had become scientific adviser to the Corning Glass Company, from which position, a year or so later, he was hounded by the Secretary of the Navy, urged on by McCarthy, Nixon, et al. The staff of the Applied Mathematics Division privately idolized Condon for what he had done for the Bureau, as they told me when they learned I was Condon's friend.

I found the AMAC group a congenial one. Many of them were interested in producing more tables of functions, now that computers could produce them faster and more accurately than before. Of course, some of the hardware enthusiasts argued that the computer rendered tables obsolete because it could calculate any function faster than one could find a table and look up the value. This was not exactly true; a program always had to be written before the function could be computed. Programming took weeks, not minutes, and there were tricks to finding the easiest and most accurate way of carrying out the calculation which had to be mastered. At any rate, there was enough belief that tables ought to be prepared, together with enough doubt as to whether tables were actually obsolete, that it appeared, for once, to be worthwhile to organize a conference to debate the matter.

It was clear that a means for producing mathematical tables was at hand, if it could be agreed that tables were needed and if a sponsor could be found to finance the work. A number of mathematicians would be willing to contribute expert advice and even some detailed assistance, if a central staff could be found to carry out the work of assembling and checking the tables as well as to coordinate the efforts and advice of the experts. The staff of the Applied Mathematics Division of the Bureau was willing and eager to take on this central task, if only enough money could be found to finance them for the eight or ten years needed to complete the job. The niggardly budget of the Bureau simply could not be stretched to provide it.

A campaign would have to be started to raise the money from some other agency. The most affluent source, the Office of Naval Research, was not appropriate; mathematical tables were not very close to the development of naval weapons. In addition, ONR was supporting many other research projects of equal or greater importance. Another possibility was the newly established National Science Foundation, rather scrawny financially just then, but specifically charged with the task of supporting pure science. I knew the first director of NSF, Alan T. Waterman, a hard-working, self-effacing assistant to Karl Compton in NDRC during the war. I wondered, when he was appointed head of NSF, whether he was going to be forthright enough to persuade Congress to provide the funds needed to support pure science in this country.

Instigation

Later, after the witch-hunt had run its course, casting the more conspicuous characters such as Condon and Oppenheimer into outer darkness, I began to suspect that Waterman may have got as much out of Congress as he possibly could have just then. NSF has never become the major support of pure science that its leading proponent, Vannevar Bush, had hoped. Too much pure research has had to be financed by the operating agencies, AEC, NASA, and the military, and thus has been subtly affected by having to pretend that its results would have an operational application. It was necessary to pretend, for example, that the measurement of nuclear reactions was desirable mainly because it would enable better atomic bombs to be designed. Nonetheless, NSF survived the McCarthy era, in part because of Waterman's low-key presentations. Unfortunately, now that a pattern has been set, it is extremely difficult to bring the NSF budget up to the level it should have to do the job originally planned for it.

The amount of money needed to produce mathematical tables was not large; NSF could fund the project if it could be shown that such tables actually would be used. What was needed was some body of experts to decide that tables were desirable. So we first persuaded NSF to underwrite a conference "on Mathematical Tables, Their Publication and Distribution, Together with a Consideration of Their Use in the Light of the Advent of High-Speed Computing Machines."

The conference was held at MIT in September 1954. Its report recommended that a *Handbook for the Occasional Computer* be put together, not only with numerical tables but also with formulas, so that it would be useful whether or not the occasional computer had a computer available. It was urged that the Bureau of Standards request NSF to supply financial aid to produce the handbook. The Bureau made the request and NSF made the grant. Two members of the NSF staff were appointed as editors, Milton Abramowitz (who died before the work was completed) and Irene Stegun. The National Research Council, at the request of NSF, set up a Committee on Revision of Mathematical Tables, which included some of those who had been most active at the conference, to supervise the work and to advise NSF of its progress. Work started in 1956, and in 1964 the U.S. Government Printing Office published the 1060-page *Handbook of Mathematical Functions, with Formulas, Graphs and Mathematical Tables,* an impressive embodiment of the conference's recommendations. It sold well from the start; in 1969 the hundred-thousandth copy was sold. I received copy number 100,002 (Allen Astin, then director of the Bureau, got copy number 100,001) in appreciation of my assistance as chairman of the conference and of the advisory committee. The whole initiating operation might seem to have been unnecessarily prolix and indirect, but I am convinced that all of the steps, taken in order, constituted the smoothest and quickest way to achieve the desired result.

The support of university computation centers and computing research was administered by a different division of NSF from the one that had helped the Bureau put together the tables. As with other divisions at that time, this one was provided with an advisory panel—in this case, on university computing—which met from time to time to assist the division chief in deciding how grant funds were to be allocated. I joined the panel in 1954 and was its chairman in the last year of my three-year term. Its operation and my relationship to it were typical of the problems involved in governmental support of university research.

The panel members came from various universities; all of them were experts in computer use or computer design. At that time, they were the only people with enough knowledge and experience to be able to give useful advice. Since NSF then had difficulty in retaining established experts as division chiefs (the computing division changed chiefs twice during my three-year term), it was the advisory panel that provided the judgment and continuity of policy.

Questions of conflict of interest could have been raised. Nearly half the panel members were directing or participating in projects funded by the division they were advising. We naturally were aware of the dangers of this situation and meticulously avoided commenting on our own projects, leaving the conference room when their funding was being discussed. But we did read all the other proposals and could model our next year's request for funds on the most successful ones we had seen.

At the time, however, there was no other way for NSF to disburse grants so as to assist most effectively the progress of university computation. No one in NSF knew the requirements and potentialities of computer science as well as those who were developing it in the universities. Later, when NSF managed to recruit and keep more experienced division chiefs, the need for experts from outside diminished and advisory panels lost much of their authority, whereupon the system developed other defects. Sometimes the division chief would ride his hobby of the moment, causing newly purchased computers to be without support for adequate staff or frustrating a group in the midst of developing an exciting idea by allowing funds to be cut off in midproject. And, as Congress asked more questions and frightened more high-level bureaucrats, the tendency grew to make grants for carefully delineated activities rather than for broad investigations. By the late sixties, it was almost necessary to describe what one expected to discover and when, and then to prove the expected discovery's practical value, before one would be granted the funds. This, unfortunately, is a habit that funding agencies are prone to. Columbus had to promise to discover the Indies before Isabella reached for her jewels.

Military consultation and committee service took up less of my time as the fifties rolled on. I remained on the board of trustees of the Rand Corporation throughout the decade, because I found much of its work interesting and most of its people stimulating. It was far enough away from the Pentagon so the gusts from political storms had little effect on its research. Reciprocally, of course, Rand had less and less effect on Air Force orthodoxy, but its research was deep and broad enough to be useful to many besides the military. I made many friends among its staff in addition to the ebullient director, Frank Collbohm.

John Williams, head of the mathematics division, was the one I enjoyed the most. His specialty was game theory, but his wide-ranging imagination shed new light on all sorts of subjects, from comments on matrix calculus to speculations on the remarkable constancy, over a span of fifty years, of traffic deaths per driver hour, in spite of the huge

Instigation

increases in the number of drivers and cars, the amount of car equipment, and the rate of speed during that time.

The Institute for Defense Analyses (IDA), originally established to administer WSEG, was another nonprofit think tank I was in on from the beginning. However, I left its board after a few years. Equating success with expansion, IDA contracted to do other things in addition to supporting WSEG. Inevitably, the other projects conflicted enough with the work of WSEG to destroy the delicate balance between civilians and officers that Hull and I had managed to achieve, and, as mentioned earlier, the form of WSEG was destroyed. In the ensuing explosion, the civilian members of WSEG had to leave the Pentagon. They moved to one of the floors housing IDA in a building a mile or so from the Pentagon. Their surroundings are more spacious and their salaries are more generous than they were in the Pentagon under civil service, but they no longer work in adjoining-office partnership with the men in uniform. WSEG is now two groups.

During the fifties my participation in the internal affairs of MIT broadened. I continued to oversee the physics graduate students, although the increasing enrollment inevitably increased formality and the number of records that had to be kept. I continued to teach courses and to supervise thesis research. I also served my turn on the many faculty committees and did a two-year stint as chairman of the faculty.

The faculty library committee interested me, although the library was a minor part of the Institute's educational activities. I had helped assemble the physics-chemistry-mathematics library in the new physics-chemistry building back in the thirties, and I also was a heavy user of the books in the humanities library. It seemed to me that librarians were less interested in the wants of the average user of the library than in the preservation of their books, whether they were used or not. But, as I came to know Vernon Tate, the director of the Institute's libraries in the fifties, I began to realize that the lack of understanding went in both directions.

Some of my assistants devised a questionnaire to be filled out by everyone coming into the science library, asking what each of them did while there. From the results we could make a few guesses as to what the library could do to satisfy their needs and preferences. But most changes cost money, and Tate was not very good at persuading the budget committee. Eventually he gave up and moved to the more congenial post of librarian at the Naval Academy.

The library study was a small part of a program slowly being formulated which would help us apply to peacetime affairs the methods of operations research developed during the war. Everywhere I looked in governmental as well as in industrial action, I saw programs initiated without consideration of their side effects and continued without finding out whether the program was really accomplishing what it was supposed to accomplish. It was clear to those who had wartime experience that, as the operations of society grew more complex and more mechanized, the techniques of O/R would be needed if we were to control the operations, rather than allowing the operations to control us.

Enthusiasm, by itself, would not persuade adminis-

Instigation

trators to give O/R a try. A sales campaign would have to be launched on several fronts. Business and governmental policymakers would have to be shown that the combination of scientific measurement and theoretical extrapolation called O/R could help them control the operations they administered. We would have to persuade a few of them to try the methods out; others would follow suit.

If, as we hoped, the idea caught on, a supply of experts, young people trained in the methods and points of view of O/R, would have to be made available to meet the expected demand for their services. A few universities should begin training these experts, to take advantage of the opportunities as they arose. This plan meant that at least part of the academic world would have to be convinced that operations research had status as a legitimate branch of applied science, with an academic content sufficient to warrant offering courses and advanced degrees in the new subject.

Selling a few industrial companies was the easier task. A number of war-trained experts took jobs in consulting firms, such as Arthur D. Little, Inc., in Cambridge. I spent a bit of time, as a consultant to the A. D. Little firm, visiting various business executives, explaining what O/R was, and indicating how the newly formed A. D. Little consulting team could enable them to try out the new techniques without first committing themselves to a permanent staff. Some of them said their companies had no problems and some said they needed no help. A few were interested, and shortly A. D. Little had to look for more O/R experts.

By 1950, a few articles about O/R had appeared in scientific journals, such as *Nature* in England and *Science* in this country. In England, the Operational Research Society of London had been formed and was publishing the *Operational Research Quarterly*. To arouse interest in this country, I went to meetings of organizations that would invite me to talk about the subject. I gave the Gibbs Memorial Lecture to the Mathematical Society, and I lectured at several meetings of the American Association for the Advancement of Science. The Social Science Research Council gave me a rough time when I told them I believed that operations research differed both in technique and in subject matter from the social sciences on the one hand and economics on the other and thus had standing as a separate subject. I seemed to convince a few of them. Other O/R veterans also were working for the same ends. We managed to penetrate the sanctum of the National Academy of Sciences or, rather, of its operating subsidiary, the National Research Council (NRC).

Horace Levinson, a mathematician with a doctoral degree who was treasurer of the Bamberger's and Macy's stores in New York and New Jersey, had heard about O/R from Arthur "Fergie" Brown, a veteran of the Navy's ORG. Between them they persuaded Marsten Morse, chairman of the NRC Division of Mathematics, to appoint a committee on operations research to write a report on the subject that might impress the academic world. The committee members came from nearly all the American wartime O/R teams, as well as from departments of mathematics, physics, economics, and engineering of a few prestigious universities.

The report of the committee came out in 1951. That same year George Kimball and I managed to get the Navy to declassify our book *Methods of Operations Research*. It was published by the MIT Press and had a surprisingly large sale, both here and abroad. Within three years, it had been translated and published in Russia and Japan.

As soon as it began to look as though O/R would become popular, a few quacks began using its name to sell their magic. It was time to establish standards and to provide outlets for exchange of information. A few of us, some from the military groups, some from the newly formed consulting teams, and some from universities, wrote a proposed constitution for an Operations Research Society of America (ORSA). In May 1952, a founding meeting was held at Arden House, the mansion overlooking the Hudson donated to Columbia University by the Harriman family for conferences of this sort. Seventy persons attended, half of them from military O/R groups, a quarter of them from universities, and a quarter from industry or industrial consulting groups. They approved our constitution, voted dues, appointed a council and an editor for the *ORSA Journal*, and elected me president for the first year. By 1960, membership had grown to about a thousand, the *Journal* was publishing nine hundred pages a year, and the Society was sponsoring a series of monographs. By 1970, the membership was over 7500, and the *Journal* was publishing 1300 pages a year.

By 1952, there was enough interest among faculty and students at MIT for the administration to appoint me as chairman of the Committee on Operations Research, to coordinate education and research in the subject. We instituted a weekly O/R seminar and started planning a summer program for people from industry. By then the

MIT Summer Program was well developed, offering conferences and special two-week courses on new aspects of science and technology for persons from industry and government. In June 1953, the O/R Committee organized the first of the fifteen-year series of summer courses in O/R, with lecturers invited from practicing O/R groups as well as from the MIT faculty. About forty people attended the first course, a few from other universities, some from government, but the majority from industry. Discussion ranged from questions as to how to form an O/R group in a company to descriptions of various mathematical models that would be useful in analyzing an operation. One such lecture, on the theory of queues, caught my research eye.

Queues arise in a wide variety of operations. Cars are held up at a toll booth, people wait in line at banks and ticket counters, partly completed equipment piles up at bottlenecks in a production line, airplanes in bad weather stack up to land or line up to take off, equipment breaks down and waits to be repaired, articles in a store's inventory wait to be sold, and so on. In each case, items arrive, wait in a queue, are "serviced," and then leave. Arrivals are seldom evenly spaced in time; they often come in bunches, between which there is sometimes a long wait. Service seldom takes the same time for each item: one man has exact change and the next a twenty-dollar bill, or one machine takes longer to repair than another. As a result, the queue fluctuates in length and service time varies. To analyze such an operation one must describe it in terms of probabilities, finding the average delay time, for instance, or the probability that the queue will have more than six items.

Instigation

A number of papers had been written about various aspects of queuing theory. The first one was written in 1909 by A. K. Erlang, an engineer working for the Copenhagen telephone company. It analyzed the fluctuations in load in an automatic telephone exchange, a question of the interaction between arriving calls and the varying lengths of telephone conversations, a typical queuing situation. Most of the later papers also dealt with particular applications, using special tricks to obtain the answers. The discussion in the summer course suggested to me an analogy with the probabilistic theory of radioactive disintegrations, which in turn suggested a more general technique of analysis of a large number of queuing problems. The one method could be used to obtain a wide variety of answers.

Queuing theory also caught the imagination of Newton Garber, one of the graduate students we had hired to assist in the summer course. I had got stuck trying to solve an equation while trying to present my new method. Garber came by the next day with the solution all worked out, together with an announcement that he believed he could solve a more general case. This more general equation was a piece of general technique I had privately decided to explore myself. But there was room for both of us, and I encouraged him to tackle it. The job was harder than either of us expected. It took Garber two years to solve it, and the work was good enough to be accepted as his doctor's thesis. To solve it he had to use some mathematical tricks I had not yet learned. In the meantime, I had found out how to avoid using his equation.

I published a few papers on queuing problems while I

was trying to work out my general method of solution. None of them were world-beaters, but they were interesting to work out, and they helped to clarify my thinking. Finally, in 1956, I decided it was time to put together all my work on queuing. The result, after three months of typing, curve-plotting, and function-calculating, was an outline of the general technique, with examples, written in a style I hoped would be read by a few executives and by most O/R professionals, not just by mathematicians.

The book, *Queues, Inventories, and Maintenance,* was the first in the new series of monographs sponsored by ORSA. Although many of the examples had been worked on before by others, here they were assembled as pieces of a tidy mathematical model with many applications. It aroused interest here and abroad and was translated and published in France. By 1970, hundreds of articles and dozens of books had been published on queuing theory. My general technique covered only a portion of the problems that can now be solved, although that portion still includes most of the important practical applications.

By 1955 there was enough interest in O/R at MIT to justify looking for money for the support of graduate students who wished to specialize in it. This time it was obvious where to go for funds; the armed services employed many O/R experts and were interested in having us train more. Tom Killian, once my fellow-student at Princeton, was the civilian head of the Office of Naval Research at the time. I outlined to him our needs, saying I thought perhaps I should go to the Army for this contract but that I came

Instigation

to him first to see what he thought. The result was two contracts, one with the Navy and one with the Army, although neither of them was for much money, only a few hundred thousand dollars each. O/R equipment costs were minuscule compared to those for, say, computer development.

Whatever difficulty we had was not in raising money; it was in the wording of the task orders in the contracts. Since the armed services were not supposed to finance education, they could not contract with MIT to educate a number of O/R experts. They had to ask for some sort of research to be carried out. It was important for us that the research requested not be too narrowly defined; if it were, it would restrict students in using the research opportunity as their thesis research. In the fifties it was still possible, although only after some argument, to be given a general task order, such as to carry out research in operations research of possible military interest. Later, in the sixties, task orders became much more specific. As a result, in order to provide students with a choice of research subjects, we had to enter into a large number of small contracts, each for a different field of research and each requiring separate reports and separate fiscal records. By 1970, nearly a third of the funds from each contract had to be spent on clerical staff to handle the red tape.

Armed with our initial contracts, I was able to ask the Institute to authorize the formation of an Operations Research Center, with offices to house the student assistants and with an administrative staff to keep track of the students' work, to monitor expenditures, and to write the necessary reports to the contracting agencies. The faculty O/R committee exercised policy control. Herbert Galliher, who had been a part-time mathematics instructor, became my full-time assistant, and the committee began choosing the research assistants for the next year and covenanting with the various departments to assure academic credit for the work the assistants would be doing.

Research in O/R was still viewed with skepticism by many faculty members, but since we now had money to support students and there were students who wanted to join the Center, there was no active opposition. Students from all over the Institute were attracted. Two of the early ones were from the physics department. One of them, mentioned earlier, studied the effects of the variability of rainfall on the water stored in the sequence of dams on the Columbia River, to see how the yearly flow from one dam to the next could be controlled so that fluctuations in the output of generated power and irrigation water could be minimized. In line with our desire to make the work realistic, we arranged for the student to confer with the Columbia River Authority and to report his findings to them. He used Whirlwind to calculate his solutions, so he was awarded a computing assistantship. This student, John Little, went to the newly formed department of operations research at Case Institute, later came back to the Sloan School of Management at MIT, and, in 1968, took over my job as director of the O/R Center.

The other physics student worked closely with an electrical engineering student who also had an O/R assistantship. After they received their doctor's degrees, they joined an O/R consulting team; a few years later they joined the faculty at Berkeley, where they have been instrumental in transforming the department of industrial

Instigation

engineering into a prestigious department of operations research.

As the fifties went on, I began to supervise more doctoral theses in O/R. Although I was a member of the physics department, other departments were willing to let me supervise the thesis research of their students and allowed the work to be approved toward a degree in their departments. Graduate courses in various aspects of the new subject were offered in a number of departments. The basic course in probability was given by George Wadsworth of the mathematics department, and a course in optimization techniques called mathematical programming was given in the School of Management. These and others were expositions of the various mathematical techniques useful in O/R.

I gave a course, listed in the physics department, that attempted to give students experience in the practical application of these techniques. I persuaded some of the administrative staff to allow a team of students to analyze the activities they directed, to see how they worked and whether they could be run more efficiently. One year the class studied the various aspects of assigning the students to classes and scheduling classrooms. Another year we analyzed library operations. All these courses, as well as the weekly seminar, could be elected by any graduate student. Any student interested in going more deeply into the subject could get advice from the faculty member of the O/R advisory committee who was in his department. If his interest continued and he showed ability, we could award him an O/R assistantship, and the appropriate member of the advisory committee would supervise his research.

Thus the O/R Center had many of the advantages of a department of operations research. In addition, it avoided the barricades a department builds around itself, which make it difficult for a student to transfer from one department to another or to divide his time among several. The Center also had some disadvantages. At first it could not admit graduate students; the prospective O/R student had to apply for admission to some regular department and thus be subject to their standards of admission. In practice, we found this was not a serious drawback; the O/R committee's representative in the department usually had a say in the admission process and could speak up for the applicant we hoped would be admitted.

The more serious handicap was that the Center had no faculty of its own; each participating faculty member was a member of some department and had been appointed and would be promoted at the pleasure of that department, not necessarily in response to the desires of the O/R Center. This meant that I, as director of the Center, had to convince the various department heads and deans that it was advantageous to their departments to have a member with interest and experience in operations research. In most cases, I managed to do so. I still feel that the advantages of our interdepartmental nature outweighed that additional work.

In the meantime, interest in O/R was growing in this country and abroad. We began to have a few students from abroad in our summer programs. A Frenchman, Jean Mothes, came over to work at the Center for six months. He had persuaded the French National Railroads to form an O/R team—or else they had decided to form one and

Instigation

had chosen him to head it; I was never clear which. At any rate, they sent him over to learn how it was done.

By 1956 there were suggestions that an international conference would be in order. The British and American O/R societies worked hard on plans. Eventually an invitation went out to interested parties all over the world to meet in Oxford in September 1957. I was invited to give the opening address. About three hundred delegates from twenty-one countries attended.

The formation of an International Federation of Operations Research Societies (IFORS) was proposed, and arrangements were made for ratification of a charter. Sir Charles Goodeve, director of the British Iron and Steel Research Institute and active in British industrial O/R, was elected the honorary secretary of the Federation-to-be, and it was agreed that the next conference would be held in 1960 in France.

The following week, I went to Paris to take part in the inaugural meeting of the Societe Francaise de Recherche Operationelle (SOFRO). My guide on the trip was the secretary of the new society, Charles Salzmann, a quiet, thoughtful young man, who uncomplainingly supplemented my almost nonexistent conversational French. It was the beginning of an enjoyable friendship and collaboration.

By 1955 the use of computers at MIT had outgrown the capacity of Whirlwind. Digital computation had won out over the older analogue computation, represented by the Bush differential analyzer. An electronic version of the differential analyzer had been built during the war, with vacuum tubes and wires replacing the gears and shafts of the original, but still having the disks and wheels. In the fifties it took up an inordinate amount of space and electric power. Although Whirlwind was overcrowded, almost no one wanted to use the much slower and less accurate differential analyzer, so it was scrapped.

Before the early fifties, nearly all digital computers had been designed and built in universities, with support from a government agency, but by 1955, commercial companies began to produce some. International Business Machines (IBM) was the biggest such firm; computer development was a natural extension of its mechanically operated, punched-card machines, although it required a completely new kind of engineer to design a digital computer. IBM began recruiting these engineers from the universities, the only place they could be found. They lured away several of the Whirlwind team; they also made arrangements to use the core memory developed by Forrester. Other experts preferred more independence than IBM offered; some of them started their own companies, and some of these became extremely successful.

Progress in programming was also being made. Writing a program in machine language was a time-consuming piece of drudgery that had to be simplified. Each step in a calculation required a multiplicity of orders to the processor, each given in the proper order, all to achieve one small part of the calculation. Much of this order-writing required no great amount of thought, but it took time to write it all out and get it read into the memory; if the programmer's attention lapsed, one single error would emasculate the whole program. Programmers resented having to spend hours grinding out programs that the machine would whiz through in a fraction of a second. So at MIT as well as at other centers, people began to suggest that perhaps the computer itself could carry out much of the drudgery. Each time a series was summed, for example, the same long sequence of orders had to be written out. Why not write a program telling the machine how to write out the sequence, how to produce its own program to sum the desired series?

The brighter programmers began to put together metaprograms called compilers, which would instruct the machine how to write its own machine-language program. The original commands, written out by the human programmer, could be couched in a form close to the usual mathematical formulations—sum this series, divide by the following product, and so on. These commands would involve many fewer instructions than the final, detailed machine-language program and would be less likely to be miswritten. These original instructions, written in an easily understood "compiler language," would be fed into the memory as though they were data to be used in some calculation. The compiler program would then act on these data to produce a program in machine language, all the multiplicity of details being generated with the slavish speed and accuracy that is the computer's sole advantage over the human. The machine-constituted program in machine language could then be used to order the computer to perform the computation originally desired. Johnny von Neumann's idea of treating instructions like data was beginning to bear unforseen fruit.

The argument over the desirability of compiler programs was intense and acrimonious. The hardware experts were horrified. It was inefficient, they protested; every computation would have to be run twice through the machine; once using the compiler to prepare a program in machine language, then running this program to direct the machine to do the actual calculation. In addition, the compiler itself, the sequence of instructions to the machine to get it to translate directions in compiler language into a program in machine language, would have to be a program in machine language. And it would have to be the great-grandfather of all programs, because it would have to forsee all the different operations used in any sort of computation and would have to guard against all the logical errors that might occur. It would take dozens of man-years of the best programmers' time to write one, and, when written, it would be so big there would not be room for anything else in the machine's memory.

The software enthusiasts had answers to all these objections. They said that computers were built to perform mental drudgery. Although in the early fifties they were few enough and expensive enough that it still might be cheaper to have a person rather than a machine write a program in machine language, in ten years it would be the other way around. In the end, why should a man do what a machine can do faster and more accurately? In addition, many people who will want to use computers will not want to go through the grind of learning machine language; computers will simply have to have enough money to accommodate a compiler if they are to be useful to everyone. True, a compiler will be a tough job to put together, but, if it is done right, it needs to be done only once for each kind of machine.

The compiler proponents soon won out. In less than ten years all large computers had compilers written for them. Indeed, as time went on, different compilers were created, as people began to realize that the machine could carry out other tasks than purely numerical calculation. Machines could sort and compare, for instance, could look up a word in a stored dictionary and copy out the definition, could in the end carry out any sort of mental drudgery. It began to appear that the electronic digital computer could be to the white-collar worker what the gasoline engine had been for the blue-collar worker.

IBM persuaded two of Whirlwind's best programmers to join their team writing a compiler for their big machines. They called the compiler language Fortran (Formula Translation). It is still one of the most-used compiler languages. IBM people were around MIT often, and I made friends with a number of them, particularly with Cuthbert Hurd, who was at the time directing some of their development work. I heard from Hurd how IBM was progressing in building and selling big machines, and he listened to my vision of getting most of MIT to learn how to use computers. I fancied that in ten years or so most of the students at the Institute would be able to use a computer as habitually as I had used a slide rule when I was a student. All of us were underestimating the speed of computer progress in those days.

It was natural for me to suggest to Hurd that it would be good business for IBM to install one of their biggest machines at the Institute. We had contributed a lot to IBM's progress; Forrester's core memory and some of the Whirlwind team had helped develop the Fortran compiler. Besides, I suggested, if students learn about computers on an IBM machine, they will be more likely to prefer IBM later, when they get into positions in which they could influence computer purchase.

I never found out how much weight these arguments had, but in 1956 Thomas J. Watson, Jr., the president of IBM, offered a machine to the Institute. The new computer, an IBM 704, was installed in the newly built Compton Laboratory in the fall of 1957. It was the most powerful machine IBM then built, with 16,000 bits (cores) of memory and all the necessary auxiliary equipment, such as tape units with their whirling reels, card readers, and high-speed printers. In those days, vacuum tubes were used for all the electronic circuits. There were hundreds of them, generating so much heat that a large air-conditioning plant had to be installed to keep the temperature under control. Tube failures were the most annoying causes of malfunction; one burnout could disrupt a big calculation and require a complete rerun. Later, when transistors replaced vacuum tubes, the much more powerful computers were not at all as impressive in size or in heat production.

At IBM's suggestion, other New England educational institutions were invited to use the equipment up to about a third of its capacity, the rest of its time being available to MIT students and faculty members. By 1960 some thirty colleges and universities were participating, their use being coordinated by a committee of institutional representatives. A number of institutions now having important computing centers of their own, such as Dartmouth and the University of Massachusetts, had their first experience in computing at the MIT Computation Center. IBM financed a dozen fellowships in computer use for MIT students and a like number for students and junior faculty members of the cooperating institutions.

I persuaded NSF and the Rockefeller Foundation to

support research in computer applications. Funds from these organizations could be used to hire postdoctoral research associates, and I endeavored to persuade interested departments to appoint these researchers as part-time faculty members, so that regular courses in computer use could be started. The mathematics department proved to be the most hesitant; many of the older members viewed the machines as mere number-crunchers. When two enthusiastic young mathematicians, John McCarthy and Marvin Minsky, joined us, they had to teach their courses at first under the aegis of the electrical engineering department. As a start in learning how a computer could solve problems rather than just calculate, they and a few students wrote programs for the machine to play checkers and chess. At first, the quality of play was amateurish, but much was discovered about how a machine could learn from its own experience.

The new Computation Center came into being just in time. The Navy had stopped financing the further development of Whirlwind, and in a few years the pioneering computer was, by comparison with the IBM 704, so inefficient that it was scrapped. Starting with Whirlwind in 1950, the computing equipment at MIT has roughly doubled in capacity every three years, having increased by a factor of more than a hundred by 1971. In spite of this, demand nearly always has outstripped supply. Each newer and faster machine has had excess capacity for about six months and then its capacity has been swamped again. Well before the ten years I had estimated, most of the students at MIT were regular users of the computer.

Research in machine use took many forms, from more

detailed and thus more realistic calculations of the internal structure of metals, of molecules, of nuclei, and of stars to the design of petroleum-refining processes and the simulation of the economic behavior of countries. McCarthy and Minsky were programming the machine to learn from its mistakes, and Victor Yngve explored the possibility of using the machine to translate from one language to another.

As might have been foreseen, the new users quickly became impatient with the Center's limitations, and we had to scurry to keep them partially satisfied. Nearly all of them used the Fortran compiler, but they complained about the work needed to find out why some of their programs gave wrong answers. In those days, the data and the program both had to be transcribed on punched cards to be handed to the operator, after which followed a wait of an hour or more—or even overnight, if the Center was particularly busy—before the answer on a printed list or another deck of cards was ready. People asked why they had to come to the Computation Center to hand in their problems and to pick up the results; why couldn't the input and output of the machine be transmitted by wire to and from their own offices?

The arguments over the possibility of remote access to the machine nearly split the Center in two. The general outlines of the problem were soon clear; the argument was how to solve it. If users were to have access to the computer from their own desks, it was inevitable that several of them would want access at the same time. This did not create an impossible situation, however. Computers act many thousands of times faster than people do. While one user was feeding in punched cards or spelling out his instructions on a teletypewriter, the machine could be busy acting on another user's instructions. When the card deck was read or the typing was completed, the machine would usually have finished the previous job and would be ready to take on the new one. It was a queuing problem, and it looked as though the average delay would be only a few seconds unless there were dozens of simultaneous users. All that was needed were the additional hardware and software to permit the machine to switch back and forth between simultaneous users without becoming confused. But that "all" was not easy.

The hardware experts felt that the switching should be done by adding new circuits and permanent connections. McCarthy and Minsky, however, were sure that a hardware solution would not be flexible enough. Hardware would be needed, of course. More memory would be required to hold the programs of all the simultaneous users in addition to the compiler program and all the other program aids. But much more software was going to be needed, to instruct the computer as to which user it should go to next, what to do with the last user's program until it was his turn again, how to bring in the compiler or other subroutine when a user asked for it, how to tell the inexpert user what to do next or what he had just done wrong, and how to feed out the answers without a delay while the processor tells a thousand-times-slower typewriter how to record the answer.

We, and a few at other computing centers, felt sure that this sharing of a machine among many simultaneous users was the next big step in getting the computer to do all sorts of drudgery. We were sure that time-sharing would become economically feasible, and we wanted to be ready when it did. Of course, a communication-system program of the sort proposed by McCarthy and Minsky would be of a size and complexity far beyond even the compiler program that had been thought impossible only a few years earlier. The limits to the program's capabilities would be the size of the computer's memory and, even more important, the dependability of the machine. With all the myriad calculations, compiler translations, and transfers from one user to another, if the machine made more than one mistake a day—one error in a billion operations—people would not trust the results and would not use the system. The vacuum tube machines of the fifties were not capable of this inhuman dependability. However, core memory units were rapidly becoming cheaper and more dependable, and the all-transistor computer promised us by the early sixties might be reliable enough. At any rate, it was time to start research.

Two projects were begun. One concentrated on the hardware approach; it was headed by Herbert Teager of the electrical engineering department. The software project was headed by Fernando Corbato, who had been a graduate student in physics supported by the Navy-Whirlwind assistantship fund. By the time he obtained his Ph.D. with a thesis on solid-state theory using machine computation, he had decided that he preferred computer science to physics, and he became my deputy at the Computation Center. He started working on the software solution, with encouragement and suggestions from McCarthy and Minsky. By the sixties the software approach had won out, although Teager had developed a useful unit that could draw graphs as directed by the computer.

10
Promotion

During the late fifties and the sixties, my activities continued to follow a number of different strands, intertwined in time but each formed by its own relationships and imperatives. One strand was my work at MIT; the teaching and the progress of the Computation Center and the Operations Research Center. Another was my participation in various advisory activities in Washington, most of them related to operations research or to computing, but going on separately from my Institute activities. Beyond these, and of growing importance, was a campaign to extend the training in and use of O/R to other countries and to more socially useful applications. To follow each of these strands separately is to produce a more smoothly flowing narrative but to lose the pattern that grew as the strands braided snugly together. Later in this chapter, we can look at the 1961-62 part of the braid.

The first strand, of course, was my work with the students. I had agreed to help David Frisch, a colleague in the physics department, in teaching a senior course, Thermodynamics and Statistical Mechanics, during the spring of 1959. The subject deals with the aggregate motions and reactions of large numbers of atoms, a beautiful and highly abstract branch of physics, having subtle connections with nearly all fields of science. I had felt a lack of facility in the subject during my calculations of the opacity of stellar material in 1939 and again later when I was investigating some aspects of plasma behavior with Will Allis.

A year later I gave the lectures myself, using someone else's text. During the year after that, in the spring of 1961, I tried out my own material, writing notes for the class and making up problems. By the end of that term I knew what I wanted to include, and I arranged with W. A. Benjamin, head of a new publishing company, to produce a paperback edition in time for use by the class in the spring of 1962. Writing the book had been a rush job, converting notes into a connected manuscript by August 1961 and reading proof on weekends during that fall. But the text, called *Thermal Physics,* was ready in time for the spring 1962 class. It was well received by the students. Their cooperation enabled me to correct many errors that had slipped by and to improve clarity in several places. I was gratified that a number of other schools adopted a hardbound edition of the text.

That four-year sequence of teaching statistical mechanics was particularly satisfying because I had always wanted to learn thermal physics and, for me, the best way to come to understand a subject is to try to get argumentative students to understand it. The physics of the effects of heat on matter in bulk is a complex subject, involving sophisticated concepts. The broad-brush treatment, thermodynamics, is usually condensed into three highly abstract "laws" that tend to be viewed as philosophical imperatives but are really the final distillation of a century of experimentation. The laws enable us to express the thermal behavior of a substance, whether it be gas, liquid, or solid, in terms of a few basic properties, such as heat capacity and density, that must be measured as functions

Promotion

of temperature and pressure for each substance. From these measured properties, by means of the laws of thermodynamics, we can predict all the other thermal idiosyncrasies of the substance.

While thermodynamics is usually sufficient for engineers and chemists, it cannot by itself satisfy the modern physicist. The physicist must relate these basic thermodynamic properties to the atomic structure of the substance; he must carry the connection between theory and experiment down to the atomic level. Only then can he begin to understand—and then predict—the amazing behavior of some substances at very low temperatures, where, for example, some metals lose all electric resistance and liquid helium loses all viscosity.

In the text and in my class, I tried to present this atomic-level part of the subject, which is called statistical mechanics, as a scientific detective story. I discussed how theories were tried, then modified to fit new measurements, and then extended to explain other phenomena and, on the other hand, how statistical mechanics, in explaining these measurements, cast new light on quantum theory itself. My attempt in the lectures was to let the student feel some of the excitement the original investigators felt, not just to contemplate the subject as a completed masterpiece, beautiful but lifeless.

My interest in the subject was heightened because I could see interconnections between statistical mechanics and operations research. Both subjects deal with phenomena having great variability, so both subjects express their theories in probabilistic terms. Some of the techniques used to analyze the motions of molecules in a gas might be useful in analyzing the motions of cars on a road, it seemed to me. And the connections can go the other way as well. I later wrote a paper pointing out that some of the methods of analysis of the random nature of arrivals to a queue could be used to describe the tendency of atoms in a liquid to bunch together or to avoid one another.

The next strand my activities followed was the Computation Center. Directing it took a fair-sized fraction of my time during the late fifties and the early sixties. Encouraged by the IBM Fellowship Program, other New England institutions participated actively in the Center's program. About thirty of them were taking part by 1960, the more active ones being Dartmouth, Brown, Tufts, and the universities of Massachusetts and of New Hampshire. The fellowships and the short courses in programming we put on for the faculty of these schools were generating nuclei of knowledgeable computer users on these campuses. These users soon brought about the establishment of their own computation centers. But even when this happened, the use of the MIT computer did not lessen, for the Institute machine could handle problems beyond the capacity of the smaller machines initially installed on other campuses.

By 1960 I had obtained grants from NSF, ONR, and the Rockefeller Institute for research on the use of computers in a variety of fields. Victor Yngve worked on a language-translation program. McCarthy and Minsky and their graduate students explored ways by which a computer could solve problems by itself; they called their project artificial intelligence. In addition, of course, many physics and engineering students used the machine in its more usual role, as a high-speed, accurate computing assistant.

Jay Forrester had completed the Whirlwind project and had moved to the School of Management. He and his students developed a computer program called Dynamo, which could simulate the internal interactions among investment, inventory, manufacturing, and sales effort of a company, to show how changes in policy might create dynamic disharmonies in operation. Later the Dynamo program was adapted to the large-scale behavior of a city or small country, relating construction, population flow, taxes, and other activities to produce a rough estimate of the future consequences of various policies. More recently, the Club of Rome has used this program to demonstrate some of the detrimental effects of the too-rapid growth of population. Of course, the realism of the results of Dynamo, or of any other simulation, depends entirely on the accuracy with which the program can be made to duplicate the input parameters and the multitude of different interactions that actually occur in any of these complex systems.

The staff of the Center was, in addition, gradually putting together a time-sharing system, to permit simultaneous uses of the machine. Staff members were trying out typewriters that could be connected to the computer by ordinary telephone lines, writing programs to translate from typewriter messages to machine language and back again, and working out ways in which the computer could distinguish between several typewriters when they were being used simultaneously. The IBM 704 did not have the capacity to do all this, so the new 709 was welcome. The various parts of the time-sharing program that had been worked on could be put together, and the whole system could be tested.

We at the Center were somewhat irritated at the lack of interest in time-sharing on the part of IBM, but we hoped that our demonstration on the new 709, although embryonic, would convince them that time-sharing could have market value. Difficulties arose, however. The 709 was not well designed for time-sharing, although it did have the needed speed and nearly had a memory of the required size. We had to program around the 709's defects, and we could not expect much help from IBM. The denouement will be reported in the next chapter.

In the meantime, at the O/R Center, research was beginning on public relations, on traffic and transport, on the allocation and scheduling problems of police and fire departments—all pieces of urban operations that we would have to understand before we could tackle even broader problems. We made contact with the traffic people at the Port of New York Authority and with local agencies, so that the problems we worked on would be real ones and so that our results could be tested in the field. We had a few setbacks. An officer from the Cambridge Fire Department spent an afternoon with us and departed full of enthusiasm about what we could do for his department. He telephoned later to say that he had been told to stay away from MIT, as it was in the bad graces of the local politicos just then.

Many of these public-operations problems needed the assistance of a computing machine to be solved, so the fact that I was responsible for both the O/R Center and the Computation Center was an advantage. The simulation of a complex operation on a computer became a popular past-time. The machine could be programmed to simulate requests for items in an inventory, for example, with all the variability in time and kind that occurs in reality. But the computer simulation would run much faster than reality, so an investigator could run through the experience of several years in a few hours. If he then added to the program the various rules for reordering and the fluctuations in the time required to replenish inventory, he could duplicate the behavior of a whole inventory system, so as to see how well the rules of operation really worked. Herbert Galliher in this way simulated the Army's huge spare-parts inventory system, from factory to warehouse to supply sergeant and from carburetor to machine screw. After analyzing the results, he suggested a few changes in operating rules that reduced the needed size of stocks by a few percent. Since the cost of the Army's inventory is several billion dollars, the saving was far more than the Army had spent, or would spend, in its support of the O/R Center.

As computer capacity expanded, many other problems were tackled. Walter Helly programmed the computer to simulate the behavior of a single line of automobile traffic, with the capabilities and variabilities of cars and drivers built into the program: reaction times, maximum accelerations, preferred headway between cars, and so on. The simulated line of cars could be started and some fluctuation in the flow could be introduced to see what would happen—whether or not a chain collision would occur, for example. Such a simulation was a speedier and safer way of experimentation than would have been possible using actual cars and drivers. Data from the Port of New York Authority on the traffic flow through its vehicular tunnels were used to verify the realism of the simulation. The computer would duplicate the buildup of "shock waves" in the flow, waves the Port Authority was trying to eliminate. Helly's simulations suggested a remedy that was tried and proved successful. Collisions were reduced, and flow was increased in heavy traffic conditions.

Still more general problems came under study. Several members of John Williams's mathematics division at Rand developed a variety of computer-based techniques for finding the optimal allocations of people (or goods) among alternative tasks. These techniques have come to be called mathematical programming (not related to computer programming). Utilizing the given limitations on supply and on production rates and the given costs of supply and the prices of end products, these techniques will indicate how much of each final product should be made to obtain the highest profit. The techniques were quickly adopted by oil companies to plan yearly purchases of various crude oils and yearly scheduling of their refineries. It was obvious that the same techniques could be used by a country to plan its use of agricultural land, for example, or to allocate its scarce technical manpower among high-priority tasks. One of our graduate students, Ronald Howard, extended the Rand work, devising an analytic tool to assist management in choosing among alternative policies.

A few years after Howard's work, another graduate student, John Jennings, used simulation to study the process of collecting, distributing, storing, and delivering whole blood in the hospitals of Massachusetts. This inventory system has an added complication: stored whole blood is not usable after twenty-one days. If every hospital hoards its supply of different kinds of blood to guard against being out of some type when demand for it comes, it inevitably tends to lose a lot by outdating. Jennings showed by simulation that a single hospital, when it balanced as best it could between occasional shortages and loss by outdating, could not do much better than most hospitals were doing at the time. However, if a number of hospitals could cooperate, either all drawing from a central store or equalizing supplies by periodic exchange, the situation could be measurably improved. Simulating different ways of cooperation and estimating costs for each, Jennings found a procedure that was not costly, was simple to operate, and did not offend each hospital's feeling of autonomy. Jennings' results have been extended and implemented by several blood programs, with corresponding reductions in outdating and shortages.

As I have said, many traffic problems that we investigated involved queues, which revived my interest in queueing theory and its applications. Automobile traffic on a two- or three-lane road, with traffic moving in two directions, is an example: cars queue up behind a slower car until a gap in the opposing traffic allows one or more to pass. The problem of understanding this behavior requires computer simulation to obtain a detailed solution but, as usual, the many details in the simulation hide the broad properties of the action, such as the sudden transition from easy passing and few queues to infrequent passing and long queues as the traffic density is increased, a situation illustrated every afternoon on roads out from a city.

Several students programmed simulations of this operation as their thesis research, assuming varying degrees of complexity in car and driver behavior. It was not until

1970 that I hit on an integral equation that had few enough parameters, yet included enough of the important characteristics of range of car speed and variability of passing opportunities, for the solution to begin to behave like real traffic. A senior, Harold Yaffe, made the observations needed to verify the theory, and we published a joint paper giving both theory and corroborative data.

The delights of research in O/R are multiple. To me the pleasure coming from understanding how traffic behaves is as great as that coming from understanding how two atoms combine. In addition, the practical applications of O/R theory are often immediate and satisfying.

Believing as I did that operations research—or systems analysis, as it was sometimes called—had the potential of becoming a powerful tool of social planning, I felt that the next step in its development had to be to extend its applications to more countries and social systems than just the United States and Great Britain. It was possible, for example, that the method might be even more successful in the less-developed countries, where the systems of operations are not so complex. However, it would be best to start in Europe. If we could enlist some able European scientists, we would have broadened our base and could see our methods first tried out in social systems not too different from ours. When a group of Europeans, financed by NATO, attended the 1958 MIT summer session on O/R, I began to wonder how I could widen the contact. It would seem to be more efficient to send four or five teachers to Europe than to invite thirty or forty students to the States. NATO was primarily military and my chief interests were in peacetime developments, but possibly these aims would not clash. It was time to pull some strings.

Promotion

I learned that Norman Ramsey, with whom I had worked at Brookhaven, was on leave from Harvard to be the NATO Assistant Secretary General for Scientific Affairs, a post that could not be entirely military. I arranged to see him in the fall of 1958, when he returned to Cambridge for a short stay. He told me his job was to control the NATO expenditures in support of research and to oversee the operation of various establishments, among them the Training Center for Experimental Aerodynamics (TCEA), just outside Brussels, and the SHAPE Air Defense Technical Center (SADTC) near the Hague. He reported directly to the Secretary General of NATO, a European civilian, and to the NATO Science Council, with scientist representatives from each NATO country. Most of the money for these NATO activities came from the U.S. Department of Defense, and any O/R work within the NATO centers would have to be approved by the Department's Mutual Weapons Development Program (MWDP). In addition to knowing Ramsey, I also knew his designated successor, Frederick Seitz, another physicist.

I found another string to pull when I learned that Bernard Koopman was in the process of getting a leave of absence from Columbia University for 1959, to manage the London liaison office of IDA. This office had been inaugurated in the early fifties, at my suggestion, to maintain coordination between WSEG and the British military O/R groups; it was loosely attached to the American Embassy. Koopman, a distinguished mathematician who had been a member of the Navy ORG during the war, agreed with me about the desirability of an expanded application of operations research. He also knew how to pull strings.

Between us, Koopman and I worked out a preliminary

program for the following summer that might appeal to the various agencies that could sponsor O/R development within NATO. We suggested a team of four or five U.S. experts who would give a two-week O/R course in Europe and then visit other NATO countries. Koopman tested responses on one of his Paris vists. The opportunity to present this program came when I was asked in December 1958 to come to Washington to see Gen. Earl Larkin, the head of MWDP. I notified Koopman, and together we went to see the general, who had heard echoes from NATO of our suggestion to send O/R experts over to Europe sometime during the summer of 1959. Koopman was able to add details of the background of the request, and I could relate our MIT experience with NATO personnel during the previous summer's O/R course.

General Larkin was a relaxed sort who knew his way around the Pentagon, although he often complained of the inconveniences of his desk-bound job. "My only service injuries are hemorrhoids," he growled. But he knew the IDA group and liked the idea of "carrying American ideas to the unconverted," as he expressed it.

Together we reached preliminary accord on the program Koopman and I had outlined. The two-week course would be given at one of the NATO technical establishments (TCEA, for example) for persons from the NATO countries chosen by the Science Council. The subsequent visits to member countries would be arranged by the Military Assistance Advisory Groups (MAAG) of the U.S. embassies in the different countries. The eight countries likely to be involved—France, Belgium, the Netherlands, Norway, Germany, Italy, Greece, and Turkey (England did not need our missionary services)—were too many for all

Promotion

the experts to visit all of them, so the group would split in two, with each half to visit four countries. Financing the experts' travel would be done thrrough a contract with MIT, which I was sure I could arrange. Other expenses would be handled by the appropriate NATO or MAAG establishment. My immediate task was to recruit the experts, to prepare a text for the course, and to write a detailed proposal that General Larkin could use to get Pentagon approval for the program as a whole. I would have to visit General Larkin's Paris office some time in the spring to ensure that all of the separate organizations would mesh.

The team of experts was soon recruited. Koopman, of course, was one. My wartime colleague George Kimball was another; he had recently left Columbia to become a vice-president of Arthur D. Little, Inc., with the assignment of expanding their O/R activities. He was glad to be able to add to his European contacts. Two MIT faculty members who had been active in our summer programs were included: George Wadsworth of the mathematics department and Herb Galliher, my administrative assistant at the O/R Center. Finally we included that bright young man, Ronald Howard, who had just completed his graduate work in record time with a doctoral thesis on dynamic programming and decision theory. For the text I could use material we had been putting together for our MIT summer programs. The formal proposal was written in a week and the Pentagon approved it.

The trip to Paris was made in April 1959. I saw dozens of people, all of whom had something to do with the program. The three days of the trip were a quick course in the complexities of an international operation. I began to sense the dichotomy within the U.S.-NATO operation. The lower-level American personnel, the majors and colonels, were efficient and accurate in arranging and reporting on the status of their part of the planned operation. On the whole, they and their families lived in American compounds, bought American supplies from the PX, and, as nearly as I could tell, seldom spoke anything but English. They were friendly enough with their European colleagues, but a tinge of paternalism betrayed their assurance that, since they disbursed the cash, they would have the final say. The colonel shepherding me about drove a Buick with skill and decision. He complained, however, that the parking spaces in Paris were too small, not that his Buick was too large.

The growing complexity of our operation worried me. I learned that more than a hundred participants were expected to attend the course and that the NATO team of simultaneous translators was going to be used, with all the attendant paraphernalia of headsets and switches to shift from English to French to German to Italian. Some of my worries were allayed when I learned that we had an ideal person to be our technical assistant. Pierre Rosentiehl had spent six months at the MIT O/R Center and had returned to France to serve out the military or other duty France exacts in return for tuition-free education at the École Polytechnique.

When the sessions opened that August, the audience numbered 124; half were from the military, some were from universities, some were from governmental agencies, and a few were from industry. We scheduled conference periods when our group of lecturers could talk individually with each of the participants.

Once the sessions were under way, I went to Paris to renew my acquaintance with Charles Salzmann and Jean Mothes, whom I had last seen at the O/R conference at Oxford. Mothes matched my early preconception of a Frenchman, voluble, gesticulating, and a gourmet. His tour of duty for his government was to run the O/R group assigned to the French National Railroads (SNCF). He introduced me to his superior, Marcel Boiteux, who was later to become head of Électricité de France and president of the French O/R society and who was to succeed me as secretary general of IFORS. As I listened to Mothes explain his group's work on train scheduling and equipment maintenance and on the application of modern electronics to signals and communication, I began to see why European railroads were so far ahead of American railroads in efficiency and service. Mothes's tour of duty with the SNCF ended a few years later, and he went on to join an O/R consulting firm and then to become the O/R consultant to a large French investment company.

Charles Salzmann was taller, more deliberate, and more single-minded. He had been in the French underground during the war. He was a Gaullist, much interested in national affairs. An admirer of Kennedy, he was also full of questions about the U.S. He had already obtained enough financial backing to set up a consulting firm, with a moderate-sized computer and a talented group that was doing operational analysis for Air France and other companies.

On Monday I reported to General Larkin on progress and then went to see Fred Seitz, who had by then taken over as science adviser to NATO. The O/R work at SADTC at the Hague was discussed, and Seitz asked whether I could suggest any further activity to follow up our present program. It was the opening I had hoped for; I said I would work up a proposal during our visits to the various countries. I did suggest that perhaps he should appoint an advisory panel or a committee to suggest developments and to coordinate his division's O/R activities. Seitz asked me to suggest members and possible duties for the panel when I sent him the proposal. He pointed out that if it were a panel, we could choose the members, whereas if it were an official committee, the NATO countries would do the choosing.

The rest of the lecture program went off smoothly. On an off day, Kimball, Koopman, and I visited SADTC at the Hague and met the group doing O/R work on air defense, which had members from the United States, Britain, Belgium, and Italy. The group indicated that they would be glad to take young people from a NATO country for work training, which suggested some ideas to recommend to Seitz. At the end of the week, after the ceremonies ending the course, Koopman, Howard, and Wadsworth started for France, Italy, Greece, and Turkey, while Kimball, Galliher, and I started for Germany and Norway (Belgium and the Netherlands had already been visited).

We had a curious two days of sessions in Germany. The group attending them was about equally divided among industrial, governmental, and academic representatives, and it appeared as though they had never before talked to one another. Every person that reported O/R activity—and some of the work reported was very good—spoke as though he were the only one in Germany doing O/R.

Promotion

During our part of the discussion, we suggested that an O/R society might improve cross-communication, but the reactions this suggestion provoked made us suspect that a single society including people from industry, government, and universities would be unthinkable; it was discouraging to us to realize that three societies might be required.

A day in Bonn spent talking to the military was not much more heartening. They listened and asked many questions, but appeared uncertain as to how O/R could be adapted to the German staff system. There seemed to be no place in the military organization chart for a civilian group, particularly one containing academicians.

Our next visit, to Oslo, both heartened and exhausted us. The Norwegian Defense Research Establishment (NDRE, directed by Finn Lied, a member of the NATO Science Council) already had an O/R group, headed by Eric Klippenberg, and there already was a Norwegian Operations Research Society, organized and run by Kirsten Nygaard. Our first day in Oslo was spent at NDRE in solid technical discussions. The second, organized by Nygaard and his assistant, Francisco Falch, nearly wore us out. In the morning we each gave a talk to the military establishment and then had lunch with the Norwegian Chiefs of Staff. In the afternoon we put on a show for a group of engineers and industrial people and, after a hearty dinner, gave still another series of talks to the O/R society. Though we were completely talked out by the end of the day, we felt we had communicated. In Norway I found few of the differences in customs and thinking that had hampered my understanding of the countries farther south. We found we laughed at the same jokes, and we felt no need to pull punches in an argument. Falch later came to MIT for a year to work at the Computation Center and then returned to Oslo to become a computer expert for the Norwegian government. He devised one of the cleverest, computer-assisted origin-destination surveys that I have seen, to measure the daily movements of the people of Oslo. Through Nygaard, who was active politically, I had a glimpse of the workings of the government of Norway, a country small enough and homogeneous enough so all people of importance know one another.

Meanwhile, wheels had been turning at NATO. The science adviser, Seitz, sent me a formal request to organize the Advisory Panel on O/R (APOR), which would report to him first on the possibilities for increased O/R activity in NATO. It seemed to me that APOR should not be large, but that it should have representation from all the European countries active or likely to be active in O/R. Koopman and I went over lists of people we knew. Sir Charles Goodeve, director of the British Iron and Steel Research Institute and at that time secretary of IFORS, seemed a good idea for the British representative, although he was more connected with industry than with the military. Charles Salzmann seemed to have the right interests and connections in France. Eric Klippenberg was close to the Norwegian member of the Science Council. For an Italian representative I nominated a professor of statistics at the University of Rome, Giuseppe Pompilj, who had attended the Brussels course and had shown considerable interest. From Germany I tentatively chose Henry Goertler, a professor of economics at the University of Freiburg. We might want Canadian and Dutch representatives later. I would be the U.S. representative; Koopman could attend meetings in his capacity as head of the London liaison office. The choice of representatives for the panel came out predominantly academic and nonmilitary, but I felt that a working knowledge of O/R was more important than close contact with military problems, which could be acquired from Seitz's staff.

The first meeting of the panel was held in January 1960 in Seitz's office in Paris. We thrashed out a proposed program to be presented to the Science Council, which was to meet two days later. We recommended organizing a sequence of summer courses and conferences on O/R subjects, sending visiting experts to lecture and consult on O/R, and financing O/R fellowships for young people from NATO countries to study O/R in universities offering such education or to get work training in establishments like SADTC.

Koopman and I then went to General Babcock, who had replaced General Larkin as head of MWDP, and got his agreement to support a few courses and conferences and to arrange a contract with IDA to finance the expert advisers. Norway and Italy had already indicated their intention to invite an advisor.

The next day I attended that part of the Science Council meeting in which our recommendations were presented. Opposition was to be expected. The council was made up of scientists in well-established fields of science: physics, mathematics, astronomy, and the like. The NATO funds they controlled had been expended in support of their own fields. Few of them had heard of O/R, some were not sure it was a science at all, others considered it to be "soft," somewhere between economics and sociology and thus not to be encouraged.

My necessarily short explanations did not convince many; their questions indicated they were not inclined to divert NATO funds from their own projects to aid this new and suspect activity. Great Britain's Sir Solly Zuckerman was sarcastic about the proposal, although he had worked in O/R during the war. The U.S. member, I. I. Rabi, wondered whether the limited funds available for the council to allocate should be taken from other projects to support this new one. Others said little; their countries did not contribute much to the budget.

At a psychologically opportune moment, Seitz forestalled the impending negative response by pointing out that MWDP had agreed to finance the summer courses and the adviser program for the first five years, that the costs of the fellowships and conferences would likely be borne by the countries that sent the fellows or hosted the conferences, and that the only additional cost to the Division of Scientific Affairs would be for an additional staff member to coordinate the program. Then the Norwegian member, Lied, came through with a strong plea for the program and the tide was turned. The proposals were approved; they would be reviewed after three years' experience. I began to breathe again, retired from the meeting, and started to plan the many things I would have to do in the next few months.

We quickly arranged for a course to meet in Freiburg the next summer. Galliher arranged it while I looked for promising advisers to go to Norway and Italy. By the time I had approval of financing and formal invitations from the host universities, it was too late to find anyone with the desired ability and reputation who could free himself for the 1960–61 academic year. Eventually I persuaded

Maurice Sasieni of the Case Institute O/R department to take a leave for 1961-62 to go to the University of Oslo, and I bullied the Rand Corporation into giving a year's leave to David Stoller to go to the University of Rome. By the time the panel next met, during the summer course at Freiburg at the end of August 1960, I could report that arrangements were definite for the succeeding year. In retrospect, I view this whole NATO exercise as a rather elementary example of international string-pulling. More complex ones come later.

The week after the Freiburg meeting, the second International Conference on Operations Research met in Aix-en-Provence. This being the first conference run by IFORS, many questions had to be settled in regard to publications, to the admission of new O/R societies, and to finances. Sir Charles Goodeve had made an excellent start as secretary, creating sensible, conservative precedents and keeping the French contingent in good humor. At Goodeve's suggestion, I was elected the next secretary, to take office in 1961 and to plan the next meeting for Oslo in 1963. Both Sasieni and Stoller were at Aix so their visits to Norway and Italy could be confirmed.

That same year I had a chance to pull two more strings. The first of these concerned Japan. During the Oxford conference in 1957 I had met Kenichi Koyonagi, one of the two Japanese delegates. He had described the interesting organization he ran, the Japanese Union of Scientists and Engineers (JUSE). In Japan, as in Germany, there was little contact between university scientists and industry. Koyonagi had set out to remedy this. His company contracted with industry for research and consultation and then hired the university scientists, part-time, to do the work. In addition, JUSE arranged to give one- and two-week courses on new technical subjects to people from industry, again using professors as the lecturers. These procedures have been fairly common in the United States since the forties, but they represented a breakthrough in Japan. At Oxford, Koyonagi had reported JUSE work related to O/R, particularly in the field of quality control, that had aroused a great deal of interest in Japanese industry. He indicated that he was going to introduce other aspects of O/R.

In January 1960, Koyonagi came through Boston on the way to England and spent an afternoon with me. He told me more about JUSE, which was getting strong industrial backing. He asked me whether he could send one or two of his staff to the MIT O/R Center for a short stay. I welcomed the idea (the Japanese came a year or so later), but then went on to tell him of our Brussels course, as an indication of other ways in which we might help.

The other string was tied to OECD. On the way to the third APOR meeting in Paris in December 1960, I stopped off in London to talk with Koopman and Goodeve. Goodeve said he had learned from his friends that the organization that had administered the Marshall Plan for Europe was to be reorganized into a permanent international body. (No international establishment can be allowed to die just because its task is finished.) The new body was to be called the Organization for Economic Cooperation and Development (OECD). It was to have a wider membership than NATO (Denmark, Austria, and Yugoslavia, for example, were in it), and its primary interest, as its name implied, was in national productivity. Goodeve felt we should get in touch with someone in

Promotion

OECD to see whether our APOR group could be of assistance.

By this time Seitz had turned the job of NATO science adviser over to William Nierenberg, another physicist, originally from Columbia but more recently from the Scripps Institute at San Diego. Koopman and I brought Bill up to date on plans and activities and arranged for the panel meeting, scheduled for the next day, to be held in his office. At that meeting the NATO O/R fellowship program was organized: a list of universities with acceptable courses in O/R was agreed on, and arrangements were made to send an announcement of the program to a variety of ministers in the NATO countries and to establish a committee to read the resulting applications and make the appointments. For the coming summer it was agreed that the course would be held in Venice and that a three-day conference would be held in Munich.

My next trip abroad in May 1961 was triple-barreled. I had first to go to Oslo to begin planning for the IFORS meeting there in 1963. Then I went back to Paris for the next APOR meeting. I was also able to visit OECD headquarters. The atmosphere at OECD was State-Department-like rather than military. A Department of Scientific Affairs was headed by a Belgian astronomer who was polite but seemed to be uncertain as to what he could do or how O/R fitted into his scope of activities. That first meeting was not very promising.

Many of these efforts at pulling strings, then so tentative and often seemingly unproductive, did in fact bring results. For example, in 1961 the Japanese organization JUSE asked for a group to come to lecture in Japan, with all expenses inside Japan borne by JUSE. A quick trip to Washington brought grudging agreement that MDWP would expand its contract with IDA to pay airfare to Japan and back for three experts if one of the courses would be given for personnel in the Japanese Defense Agency. One of Koyonagi's men came to Cambridge to work out details.

The program was planned to start in August 1961, after the NATO course in Venice that began at the end of July. Two recruits for the Japan trip were not hard to find. Ronald Howard, who had taken part in the Brussels course and had since become an assistant professor in the MIT electrical engineering department, was glad to take part. The other lecturer was James Dobbie, who had been with me in the Navy ORG during the war, had recently come to work for the O/R consulting team at Arthur D. Little, and had taken part several times in our MIT summer programs.

In Venice, I met with Pompilj and Stoller to discuss plans for the course there and for Stoller's year-long stay. Pompilj, a quiet person, proved to be highly successful in organizing conferences and courses in Italy. He seemed to know all the people who could make places available and things happen—not easy in Italy. The course about to begin was to be held in the impressive monastery buildings on the island of San Giorgio, just opposite the Ducal Palace.

Pompilj was head of the Institute of Statistics at the University of Rome. I once asked him how he happened to become interested in statistics. He grinned and said he had been in the Italian army in North Africa and had been captured when the Italian push toward Egypt was repulsed. In the British prisoner-of-war camp in India, he was so bored that he decided to study mathematics. The books in the camp's library interested him in statistics; when he returned to Italy he took his degree in it and eventually succeeded to his professor's chair.

After seeing the Venice course well started, I headed for Tokyo, where I was met by Koyonagi, carrying a mysterious package wrapped in green cloth, which turned out to contain the expense money for the three Americans for our whole stay—more than a million yen, a substantial bundle of bills.

The next two days were spent with the people who had organized the three courses. Here I met Col. Kazuo Tada, then secretary of the General Staff, who had organized the course for the Japanese Defense Agency. We three took to him immediately. He was short, muscular, completely unself-conscious, piecing out his fractured English with expressive gestures. He was very interested in O/R and had evidently worked hard to make the military course a success. He accompanied us on our weekend sightseeing trips and attended many of our lectures to the industrial people.

I once asked him how he became interested in subjects such as O/R. The tale took a lot of gesturing but it was fairly simple. During the early part of the war he had been captured by the British and sent to a prisoner-of-war camp in India. Being bored, he decided to study mathematics, and the more interesting mathematics books in the camp library happened to be on probability and statistics. (The implausible coincidences of Tada's and Pompilj's experiences can only demonstrate that probabilities have no real meaning for individual human events.)

When the Japanese edition of Kimball's and my book

on O/R came out, Tada appreciated the probabilistic basis of many of its examples and set about persuading the Japanese Defense Agency to organize O/R teams. When he retired from the army, he became the head of a firm that provides computing services and systems analysis to industry.

Through Kenichi Koyonagi I was able, during that stay in Japan, to glimpse some of the deep divisions still present between classes. Another deep division in Japanese society was that between generations. It was as yet unthinkable for young men to hold positions of authority. Our talks with younger technical men participating in the second course in Tokyo showed us that some, at least, understood how O/R could be of great assistance in making governmental and industrial policy. But these same younger men saw little chance that their ideas would be listened to. As a result, most of their O/R work was on theoretical details, divorced from practical considerations. The members of the Japanese O/R Society felt it necessary, in order to make the society respectable, to elect as president of the society a successful senior businessman who thought O/R an interesting academic exercise.

We had not, of course, expected an overnight conversion of Japanese management. The younger men were eager, and we hoped that some of them would reach positions where they could begin to apply some of the methods. The third course, given at Hakone, was for junior executives; we hoped to persuade them to give O/R a chance. Several of the attendees were executives of the Japan National Railways, and we did hear, a year later, that the railway had established an O/R group that included two of our students from the second course.

Promotion

Our lectures were not accompanied by simultaneous translators; each speaker had to proceed by fits and starts, speaking a paragraph in English, then waiting while the translator rendered the paragraph into Japanese. Two of the translators, Takehiko Matsuda and Shigeru Idei, were faculty members in the economics and management departments of Waseda University, one of the largest private universities in Tokyo. We could hope they would carry back some of our ideas into their regular classes. As a matter of fact, Professor Matsuda did take an active part in the development of the Japanese O/R Society and, later, in IFORS activities.

On our last evening in Japan we spent the remainder of our bundle of yen on a banquet for our hosts and co-workers. After the meal I tried to say how their country, as well as their hospitality, had impressed me. And then, to lighten the sentiment, I said there was one celebrated sight I had not been allowed to see, Fujiyama. "I know," I said, "that you have explained that the air in August is too misty for us to see Fuji-san from Tokyo. However, I was also taken around the rim drive at Hakone and people pointed to where Fuji ought to be, but there was nothing but mist. In fact," I concluded, "I'm not sure there is a Fuji."

As the party ended, I looked around for Colonel Tada. Finally I saw him coming back into the room, grinning. He rushed up and blurted out, "You'll see Fuji tomorrow; I just arranged to have the Air Force helicopter to take you up."

The next morning early, at an open field in one of the parks, we watched the only helicopter then belonging to the Japan Air Force coming down to land. It was an antiquated relic of the U.S. Navy, but it flew, expertly operated by two cheerful friends of Colonel Tada. There was no trouble about seeing out; both sides of the main compartment were open, with only canvas strips across them to hold onto if you wanted to lean out. As soon as we were above the smog, the air was sparklingly clear—and there was Fuji-san. It was a most satisfying termination of our Japanese experience.

That fall I shared with Feshbach the lectures in Methods of Theoretical Physics and had my usual enjoyable time guiding student research. The first stint of lecturing was finished by the end of October, when I could leave the class to Feshbach and take the plane again, this time to Athens and then back to Rome for an APOR meeting. At Athens I participated in a short NATO conference on O/R and made arrangements with the authorities for an O/R consultant to come to Greece a year or so later. The organizer of this, and the founder of the Hellenic O/R Society a few years later, was General Spannyonakis. He was soft-spoken and not at all my idea of a general, but his quiet persistence certainly got O/R started in Greece. I spent more time with Spannyonakis two years later, when I came back to help inaugurate the Hellenic O/R Society.

In all these peregrinations, I found I was continually dealing with old friends, nearly all of them physicists. Throughout the sixties I never passed through the London airport without greeting or being greeted by a friend or former colleague or student. In nearly all of the many agencies and institutions I visited I would be greeted by a schoolmate from the twenties or a comrade from the war years or someone who had taken a course I had given years before. For example, Hilliard Roderick, new head of the OECD Directorate of Scientific Affairs, turned out to have been a student at MIT, and his new assistant, C. A. Cochrane, turned out to have worked in a British radar laboratory I had visited during the war. And again, at my first meeting of the Council on Library Resources in November 1961, the first to welcome me aboard was Joe Morris, my roommate in the Graduate College at Princeton. Clearly, my ability to get things started depended to some extent on the fact that the world's scientific fraternity—especially the band of physicists—was small enough so that a large fraction of them were friends, or friends of friends, of mine.

During the fall of 1961 the Computation Center was again outgrowing its computer. The IBM 709, installed in 1960, already was being used to its full capacity, round the clock and on weekends; it had not proved versatile enough or dependable enough to support a satisfactory time-sharing operation. We hoped that the new machine, a 7090, would have enough more memory and enough fewer breakdowns so we could "go public" with a half-dozen remote typewriter consoles. It met our expectations; and within three years its successor, the 7094, would be handling more than thirty simultaneous users. IBM, however, still displayed little interest in our efforts.

By this time, most universities were struggling with the same problem we at MIT had been trying to solve, insufficient funds to satisfy the rapidly increasing demand for computer time. I worked on the problem with J. Barkley Rosser, a mathematician long interested in computing, who had been a fellow-member of the Bureau of Standards AMAC and the NSF panel. We used the ploy of a Conference on University Computing to persuade the National Academy of Sciences to set up a panel to make recommendations as to how universities could be provided with the computers they needed.

The panel struggled with the problem for several years, gathering data on costs and needs. By the time its report came out, the government was beginning to be less bountiful in financing university expansion, so only a few of its recommendations were implemented. Not every college and university in the country would be able to have its own large computer. Someone would have to work out an alternative solution to the academic need for high-speed computation.

During the spring of 1962 I again gave the lectures in Thermodynamics and Statistical Mechanics, using my text, *Thermal Physics*. My class schedule cut down on my chances for long trips out of town, although I managed to get to Washington regularly, to give an out-of-town lecture about once a month, and to squeeze into a long weekend in March a meeting of APOR in London and a session in Oslo, where we planned the 1963 IFORS Conference. The Conference was set for the first week in July, when the delegates could stay in the new dormitories at the University of Oslo and we could use the new auditoriums there. Several new O/R societies had joined IFORS during the previous year, raising the total to seventeen national societies. It began to look as though the attendance at this conference might be as high as five hundred, if we counted those from Russia, Poland, and other countries that did not yet have a society but who had been invited to send observers. Our estimates turned out to be accurate, and the Oslo Conference turned out to be a great success.

The next trip was to Mexico City, in spring vacation. Manuel Sandoval Vallarta, who had done his research in cosmic rays with Abbé Lemaitre at MIT in the thirties, had gone back to Mexico during the war to head the government agency that eventually became the Atomic Energy Commission of Mexico. Manuel arranged with the Fulbright Cultural Exchange Agency in Mexico to invite me to come to Mexico City to give a public lecture on operations research.

Another important, although short, strand of activity commenced that year. In January 1962, President Kennedy requested the National Academy of Sciences to "evaluate and recommend research on behalf of the conservation and development of America's natural resources." The only way the Academy could act on such a request was to appoint a committee and ask it to write a report. A committee of experts was appointed: Detlev Bronk, then head of the Rockefeller Institute; Dean Frasché, an expert on ore prospecting; King Hubbert, an authority on petroleum resources; Frank Notestein of the Population Council; Roger Revelle, an oceanographer and conservationist, and Paul Weiss, a biologist. I was supposed to contribute ideas on the systems analysis that would be required to knit the proposed projects into a coordinated program for future governmental action.

I hesitated to accept the appointment for several reasons. First, I was not sure I could contribute much—I did not know enough about the system linkages. I had also become skeptical about the value of reports that educate no one but the writers—the Rosser Report on university computer needs would be an example. However, the study was requested by a President who might read the report and just might act on it, so I joined. I would advance my education, at any rate.

The committee met several times during the spring of 1962 and then spent the month of July at the Academy's house in Woods Hole, Massachusetts, putting the report together. Much of what it said would make important reading today, but, of course, no one reads a report that has been in the file drawer more than a year or so. The introductory statement outlined a few realities that are only now beginning to be understood generally. Its first paragraph reads,

> The United States now possesses or has access to the gross quantities of energy, food, fiber, minerals, water and space required to maintain and even improve its standard of living for perhaps four to five decades, *if* the world situation were to remain relatively stable. But the world situation is *not* stable; other countries may preempt what we now import and new demands may exhaust our own resources more rapidly than we now foresee. We must acquire understanding of our alternatives and take concerted action if we are to maintain our present access without future denial to our children of the freedom of action we now enjoy. We live in a critical period when clear definition of the natural resources research objectives and the development plans for realizing them are of great importance to the future welfare and stability of mankind. Today the United States is able to enjoy a certain com-

Promotion

placency concerning its short-range situation; but it faces a wider arena and a longer term in which complacency would be tragically dangerous.

The report listed ten recommendations that apply now with still greater urgency:
1. Extend applied research to increase productivity in a wider range of agricultural environments.
2. Conduct basic research in plant and animal genetics and breeding.
3. Support research leading to low-cost sources of industrial energy.
4. Increase support of research in the movement and quality of surface and ground water.
5. Develop analytic techniques for the planning and management of water resources.
6. Increase support for research relative to the discovery and development of mineral deposits.
7. Conduct research on pollution and its effects on man's total environment.
8. Support ocean fisheries research.
9. Develop systems analysis capability for resource planning and management.
10. Establish a Central Natural Resources Group within the federal government, to coordinate, support, and extend this planning and research.

Each recommendation was backed up by a section of detailed proposals, tables, and graphs. For example, King Hubbert's section on petroleum resources predicted a crisis in petroleum supply by 1975 unless the U.S. were to take appropriate measures in advance. In 1963 his prophecy was severely criticized as needlessly alarmist, but we now see that his prognosis was remarkably accurate— if anything, overestimating our maneuvering time by two years.

Unfortunately, by the time the report went through the processes of review, publication, and dissemination to Congress and the executive offices, President Kennedy had been assassinated and this country had begun a decade of tampering with world stability that reduced the period of grace estimated in our report from four or five decades to one or two. Our recommendations still hold, but now the country will have to implement them in an atmosphere of crisis instead of one of unflustered determination.

Aside from a long weekend spent with the O/R department at Berkeley on the way to my last attendance at the Rand board of trustees, my only other trips out of town that spring of 1962 were one-day runs to New York or Washington. I chaired a session on transportation at an ORSA meeting and conducted an NSF-sponsored press conference on computers and physics. Additional meetings of the Rosser Committee on university computers were held. The working session of the Academy committee on natural resources at Woods Hole took up all of July. I finally managed to round out that sample year 1961–62 by hiking from one Appalachian Mountain Club hut to another in the White Mountains, with my daughter Annabella and her new husband, Hugh Fowler, in a week of alternate fog and sunshine. Surprisingly, Hugh liked it.

11
Rumination

The year 1961-62 saw the peak of my missionary efforts on behalf of operations research. Able leaders were emerging in many countries to carry on the work of applying O/R to the world's problems. Whether or not the name "operations research" survived, the method and the point of view were established; the concept and practice of using the quantitative methods of physical science to help understand man's activities would be an important part of the emerging science of social behavior, whatever its various portions might come to be called.

Of course, there was further work for me to do. I could anticipate retiring from the chairmanship of APOR in a year or two, but there was still the work for OECD. That international organization dealt with the nonmilitary activities that had always attracted most of my interest. I had started working with NATO only because I had some leverage there: I was known and respected by the U.S. Defense Department, which worked through NATO, whereas the agency supporting OECD, the State Department, seemed either to ignore me or regard me with suspicion.

By 1962, however, I had European help in establishing myself with OECD; I did not need strong support from Washington. This was just as well. The State Department still showed the devastation wrought by Senator McCarthy a dozen years before, when many intelligent and outspoken professionals had left or had been hounded out. Many of those remaining were reluctant to report unpalatable truths about the countries to which they were assigned. We still are suffering from that lobotomy. More realistic reporting by our diplomats certainly would have kept us from being caught unaware, time after time, by erupting crises.

At the suggestion of friends in and out of OECD, I had several talks with the Science Attaché and other members of the U.S. Embassy Staff in Paris. I was seeing activity in Europe that people in the United States ought to learn about, and I knew of many U.S. projects that the Europeans would want to hear about. But the idea that my program might be a two-way street did not seem to be comprehended by the State Department people I talked to in Paris and Washington. To them, the United States knew everything. The idea that some Americans should be sent to European conferences to learn somehow horrified them; the only experts we needed to send abroad were those who would tell the rest of the world how it was done.

It was well, of course, that my infiltration of OECD occurred in 1962, while goodwill toward America was still in existence. By 1965 and after, our actions in Vietnam and elsewhere had so disillusioned most intelligent persons the world over that it was increasingly difficult to get their cooperation in anything that might be U.S.-dominated. Though my friends were still personal friends, it became impossible for me, in Japan or in Europe, to have the frank discussions of politics and world affairs that I had had in the early sixties. Hesitation on the part of foreign friends about hurting my feelings and dislike on my part of either condemning or apologizing for my country's actions inhibited whole areas of discourse that had once been so interesting.

Luckily, in 1962 I could work directly on OECD projects with my British and French friends, rather than through Washington. I became a consultant to the Division of Scientific Affairs, paid directly by them. Alec Cochrane and I planned to appoint a group of experts (rather than a formal advisory panel, which would have required clearance from above) to assist our program. Cochrane, who came to OECD in 1962, was a great help in showing me the differences in structure and procedures between NATO and OECD. He was a veteran of the British civil service, the first I had known well, and his knowledge of European international customs gave me an insight that differed greatly from that I got from the U.S. Embassy people. In the following five years Alec and I were together often, planning and then running conferences or else just driving around to see the country.

Between us we picked a congenial group of experts: Jean Mothes, a long-time friend of mine, by then a partner in SEMA, the French O/R consulting firm; B. S. Barberi, a friend of Pompilj's from the Italian government's productivity agency; Gunnar Dannerstadt, who had studied O/R at MIT and had returned to Sweden to do O/R consulting; and Stafford Beer, whose British O/R consulting firm would later join SEMA to form METRA. We met initially during the IFORS conference in Oslo, but it took a year of fumbling on our part and of hard work on Cochrane's part to determine just what OECD could do to further the use of O/R in the departments of the various member governments. Finally, at a meeting in Paris in May 1964, it was decided to organize a series of conferences for the ministerial staffs of the member governments to report on O/R methods then in use in government and to hear from experts what further use could be foreseen. This two-way exchange, we hoped, would induce the participating ministers to look into the whole subject of O/R and, perhaps, persuade them to make more use of their own nation's O/R experts.

The first formal step was to ask if any of the member nations would agree to sponsor such a conference on some subject of particular national interest. Of course, the inquiry was not sent out until a few countries had already informally indicated their interest. In the end, four countries responded affirmatively: Ireland, Sweden, Norway, and Italy. During the next two years Cochrane and I visited each country to plan the conference schedules, to arrange for the participants, and to agree on the experts to lead the discussions.

By then, experts were turning up all over. Sweden had proposed a conference on O/R in hospital services, and Cochrane located Professor R. A. Revans, who had been improving the operation of a number of hospitals in London. When we visited Stockholm, we talked to two young staff members in the ministry of education who had never heard of O/R but who described to us some very interesting measurements they were making on the relation between teaching methods and rates of learning in primary schools. We told them they were doing O/R without knowing it and arranged for them to report their work at the Norwegian conference on O/R in education.

When we visited Amsterdam, we discovered a small group in the department of transport that had made detailed analyses of the daily trips of people in Amsterdam and were using computer simulation to predict traffic trends, work that was of great assistance in planning a balance between public transport and road construction. We got W. ter Hart, the head of this group, to speak at the conference in Rome on town and regional planning. At the Rome conference, the French organization SEMA also contributed reports on some excellent regional planning it had done for the French government. One was on the balance—and interference—between tourism, manufacturing, and fishing in the future development of Brittany and the appropriate budget allocations that would be required by the alternative policy decisions adjusting the balance.

These discoveries increased my annoyance with our State Department, which would finance the travel of a few Americans to tell what was being done in the United States, but was not interested in sending anyone to learn what was being done in Europe.

The conferences Cochrane and I arranged were a departure for OECD. Most of the previous ones organized by the Division of Scientific Affairs discussed how each member country allocated its scientific budget. Our innovation impressed Roderick, who became head of that division in January 1966, a step up from his previous assignment as U.S. attaché to UNESCO. After a number of discussions with Cochrane and with me and, I inferred, a number of arguments with Washington and with his superiors in OECD, he decided to support what we had started —indeed, to continue with a whole series of conferences on the applications of science to all aspects of governmental planning. I contributed to a pair of these latter conferences in London and Paris in 1967-68, on the use of computer simulation in city planning, and I spoke at a conference in Oslo in water management in 1969. After that I felt I could bow out of OECD activities.

Long before that, I had left the NATO panel, which was running well. Four conferences were held during the summer of 1964: in Toulon, in London, in Athens, and at NATO headquarters. Eric Klippenberg of Norway took over as panel chairman, and my last attendance was at a meeting at the Hague in September 1964. I did give the opening paper at a later conference on queueing theory, held in Lisbon in September 1965, just before the OECD conference in Dublin. And in March 1970, I went back to be an exhibit at the tenth annual session of the panel, held at Brussels, the new home of NATO.

I made a few other barnstorming trips. Koyonagi had asked whether I could bring another team to Japan to repeat our earlier program. Kanpur Institute of Technology in India, which has exchange arrangements with MIT, evinced interest in O/R. In addition, Col. Itzak Jacob, a former student at the O/R Center who had returned to Israel, wrote to ask whether I could help him in setting up an O/R group for the Israeli armed forces.

The Defense Department was unlikely to finance this more extended trip, even if expenses within each country were to be borne by the hosts. After some cautious reconnaissance we found that the Ford Foundation was interested in supporting our mission. The planning was additionally complicated by a request from Taiwan that came in after everything else had been settled. The expedition was interesting and fatiguing. In Japan, as before, we were stimulated by the people and charmed by the country. We came away feeling that O/R had made some progress in Japan. More of the older men took it seriously, and more of the younger men were working on actual applications.

The stay in India was both educational and depressing. The series of lectures at Kanpur Institute was given to a mixed group of about eighty students, professors from other colleges, and staff officers from the Indian Defense Department. We delivered a varied mix of lectures and hoped that all the participants got something from them. Some of the attendees reported what they had been doing in O/R. Nearly all of their work was theoretical, queuing theory being the most popular subject; very little concerned actual application. We never had the opportunity to talk with administrators, as we had in Japan, but it was apparent that O/R was not yet in use in India.

The stay in Israel was energy-demanding. We lectured at the Weizmann Institute and at the universities of Haifa and Jerusalem, where O/R courses were taught. We spent time with the Israeli Chiefs of Staff, and we spent most of a day with the O/R team organized by Colonel Jacob. The group members were working hard on what seemed to us to be the right problems. We heard later that their studies had contributed to the planning of the Seven-Day War.

By 1968 I would have to retire from all administrative jobs at MIT, although I could carry on, half-time, teaching and supervising student research for another five years, until I was seventy. In 1965 I turned over the physics graduate student office to George Koster, who had been working with me for several years. The transfer of the Computation Center and O/R Center would take more arranging.

By early 1962, the time-sharing system developed by Fernando Corbato for the Computation Center was beginning to show what was possible in making a large computer available to many simultaneous users. Others were beginning to see what a few of us had glimpsed earlier, that time-sharing opened up a whole new spectrum of ways by which the computer could assist mankind in research, in education, and in communication. Others were working in other universities on the intricate problems of increasing the number of input stations and of improving dependability and ease of operation. Only the computer manufacturers seemed to be skeptical.

Although the MIT time-sharing system was working in a rudimentary way on the Center's IBM 7090, it was obvious to us that even this advanced machine's design was not well suited for this new task. We tried to persuade the IBM people to make changes in their next computer model that would make it easier to adapt to time-sharing, but we had little success. The machine that IBM installed in the Center the following year, a 7094, had none of the modifications we had suggested, although it did enable us to link up twenty to thirty simultaneous users. If real progress was to be made in developing time-sharing, we would have to break away from our dependence on IBM.

Another consideration entered. In the late fifties the Computation Center could both carry out research in computer use and provide computer service for MIT and the other cooperating institutions. Though the primary task of

Rumination

the Center was service, the Center at first was also the only place that could do the research. The early time-sharing development could not have been carried out anywhere else in the Institute nor, indeed, at any but one or two other places in the country. But by 1962 many others at MIT, besides the staff of the Center, had the knowledge and interest to carry out computer research, and other sources of financial support for computer research, besides IBM, came into existence. In particular, the newly formed Advanced Research Projects Agency (ARPA) of the Department of Defense began to see the possible military uses of time-sharing and indicated a willingness to provide considerably more funds than IBM was providing.

Obviously some reorganization was needed. The Center's primary job was to provide service, which it could do with the IBM machines, although the time-shared part of the service was quite limited. The research and development task had evidently outgrown its role of junior partner; it should have its own separate organization and equipment. Computer science was now a major undertaking. A committee of deans and vice-presidents was formed to decide what MIT was to do. The result was a contract with ARPA and a new project, called MAC (Man and Computer, or Machine-Aided Cognition), directed for the first five years by Robert Fano, a professor of communications in the electrical engineering department. To begin, MAC used the Center's machine via time-sharing, but it gradually took over most of the research that had been carried on at the Center, in addition to starting new research. Eventually, most of the Center staff members who had been active in research went to MAC, Corbato among them. They bought an IBM 7094 with our time-sharing modifications to use until they could devise a more thoroughgoing time-sharing system, to be used with a machine designed more appropriately.

Although I regretted the dichotomy, I agreed with its necessity and did everything I could to make the transition a smooth one. But the change meant to me that it was time for me to leave, as soon as the Center had settled down to its purely service task. My initial interest in computers had been for their use in scientific research; I had never contributed to research on computers themselves.

But stability was not quickly reached. We all had tried to warn IBM of what was coming, but their lack of interest in time-sharing seemed to blind them. When the inevitable happened and Project MAC chose a General Electric computer for its next machine, it was the Computation Center that bore the brunt of IBM's shock and anger. It took the combined efforts of MIT President Howard Johnson and of Gordon Brown, the dean of engineering, to keep IBM from entirely withdrawing its contribution to the Computation Center. It was not until the end of 1966 that things settled down to relative calm again, and I was allowed to leave the director's job.

By the early seventies, my contacts with the computer world had been reduced to two. About the time I left the Computation Center, I accepted an invitation to join the board of directors of the Control Data Corporation, a manufacturer of large computers and a smaller competitor of IBM. It has been educational to watch, from the top, the operation of an industrial company in a highly competitive field, a kind of activity I had refused to enter, at the

Rumination

bottom, in my youth. I now believe I understand a bit more of the subtle pressures, governmental and social, that force a company to take action, at times almost against its own desires. It seems that even fairly large companies have little more initiative in the face of the forces of world economics than does the individual—indeed, in some respects they have even less. If I had subscribed to the theory that the country is run by a few malefactors of great wealth, my involvement with Control Data would have dissuaded me.

I did gingerly try suggesting a few changes. One of them, of course, was that time-sharing might be profitable to develop; I also indicated ways of improving the contacts between university computing centers and the CDC staff. It was illuminating to see how these suggestions changed as they moved down through the large organization, until I could hardly recognize the final result. Luckily for my peace of mind, I could persuade myself that these final results were an improvement over my initial suggestions. It is always a chancy affair to start a complex organization in a new direction.

The second continuation of my contact with computer activities was with the group of New England colleges and universities that had been organized in 1957 to use the MIT computer. As time went on, many of these cooperating institutions developed their own computing centers, although they still made use of the MIT equipment. But when Project MAC was formed and IBM reorganized its support of the Computation Center, free time for these institutions was no longer available. It seemed to me that this confederation of academic computing centers, already accustomed to cooperation, was too valuable a resource to let disband. I persuaded NSF to grant it some money, to see whether it could be transformed into a continuing mutual-assistance partnership. NSF was willing to support the effort, for it could see that a mutual sharing of computer capacity was the only solution to the college's need for computing; there simply was not enough money to buy a large machine for every campus.

The group therefore cut itself loose from MIT and formed a nonprofit corporation called NerComp (New England Regional Computing Program), with a governing board of elected representatives from the more active member schools. Over the past several years, time-sharing has developed sufficiently for NerComp to set up a rudimentary network, linking by leased telephone lines the various machines on the various campuses and connecting this net to typewriter consoles in other schools with no adequate computer of their own.

Thus, by 1970, a user at a console at Wellesley College, for example, could call up the central switchboard and be connected to any one of a half-dozen university computing centers and use its programs. The cost was little more than it would have been if the user had been at the center, and the saving in time and travel was considerable. Each of the contributing centers had a different machine, and each offered special programs the others did not have, which complicated NerComp's job of getting the net to work. But in the end everyone had more computing capacity than before, the contributing centers sold more computing time, and the small colleges no longer felt restricted in their work if they did not have a large machine.

The network had its limitations, and breakdowns did

occur, although at least half the delays were due to faulty telephone lines. In fact, our simple hookup was looked on with scorn by those who were busy designing high-powered (and expensive) computer nets. Our net, however, was actually operating, people were finding it useful, and more schools were asking to join. We were learning how to solve operational problems, which usually are more difficult than the initial design problems. By 1973, Ner Comp was operating in the black; our service charges were paying the costs of operation, and the member dues and the NSF support could be devoted to improving and diversifying our kinds of service. By 1975, it was the only self-supporting network in the world, linking a number of different machines to give cooperative service to a large number of independent public and private institutions of higher learning.

Turning over the O/R Center was easier. A number of capable young men, products of the Center and members of its staff (as well as of various Institute departments), were capable of taking over from me. Discussions with the deans of engineering and of the School of Management led to the choice of John Little, one of the first of the Institute's output of O/R-trained men, as my successor. He took over in 1968.

I still had the urge to continue exploring in several fields. Toward the middle of the sixties, acoustics attracted me again. New developments had taken place, and I believed that what I had learned while writing *Methods of Theoretical Physics* could be used in calculating the behavior of sound. I had been talking to a colleague, K. Uno Ingard, about his work in acoustics, and when I suggested that we collaborate on a new and more ambitious volume than my earlier *Vibration and Sound,* he agreed. Uno had come to the Institute's Acoustics Laboratory from Sweden shortly after World War II, and, after Bolt and Beranek had left in the late fifties to organize their own consulting company, he was the only member of the physics department still active in acoustics.

I found acoustical theory to be intellectually exciting again. Ingard pointed out many newly encountered acoustics problems to which techniques Feshbach and I had worked out for the *Theoretical Physics* volumes could be applied. I could settle down, as of old, to spend weekends and summer months sweating over equations and then struggling to explain the solutions clearly in words. Ingard contributed valuable chapters on the effects of turbulence and other motions of the medium carrying the sound waves. There was nothing earth-shakingly new in our product, but in about a thousand pages we did put together the results of much new research, together with a cement of our own work, achieving, we hoped, a harmonious and understandable pattern. The book, called *Theoretical Acoustics,* came out in 1968 and was well received.

Library operations also caught my interest again, in part because of my membership on the board of the Council on Library Resources. I could see more clearly the librarian's dilemma: too many books and too little money. I was sure the methods of O/R could help librarians make their ever more difficult decisions. Of course, these methods used mathematics, and librarians were known to be allergic to mathematics. A few librarians and some computer experts were beginning to talk about using computers in libraries, to speed up book ordering and cataloging as well as carrying out the work of recording book use.

It seemed to me it was time to write a book that spoke to both librarian and computer expert, that pointed out to the computer expert the data the librarian would need in order to balance costs against benefits, and that indicated to the librarian what he could do with the data if he had them. Writing such a book meant, among other things, working out a probabilistic model of the way the library material was used and then presenting it in simple terms. It also meant extending the model far enough so that predictions of future library use could be made, and thus the librarian could take appropriate action in advance.

This book, called *Library Effectiveness,* also came out in 1968. It had some effect. A few young faculty members in library schools are applying its ideas, extending the theory, and including the concepts of O/R in their library-science courses. The O/R fraternity also has begun to take an interest in library operations. In fact, *Library Effectiveness* was awarded the 1968 Lanchester Prize for the "best publication in English in 1968 on an operations research subject."

During the sixties my work in O/R had been taking me farther and farther from the sort of physics that I had worked on in the thirties. Our summer O/R sessions concentrated more and more on the public sector, and the attendees began to be town managers and other public officials, rather than industrial executives or O/R technicians. The Center put together several books on O/R in public systems. I still thought of myself as a physicist, but I began to wonder whether my physicist friends so considered me. So, when I learned in 1970 that I had been elected vice-president-elect of the American Physical Society (APS), I was both surprised and pleased.

The office of vice-president-elect is the first rung of a five-year assignment; by convention, the vice-president-elect becomes the vice-president and then the president, in successive years, and then sits on the Society Council for another two years. My experience in these roles was, as usual, educational. By 1970, the APS had become a very different organization from the cozy band of devotees I had joined in the twenties. By 1970, the membership numbered over 20,000, too large for its major meetings to fit into a university campus or into anything but a major convention center. The self-confident view of life of most of its members in the fifties and early sixties had been rudely shaken by the recent cutbacks of federal research funds. Employment for its younger members was hard to find, and many older ones were wondering how long their jobs would last. Anti-war demonstrations occurred at APS meetings, and antagonism between those who wanted the Society to take stands on social questions and those who wanted it to remain a purely scientific forum was barely repressed.

To me, part of the trouble was that most physicists today tend to work in a much more narrow range of specialties than we had done back in the times of the great Depression. Then, we had considered that any quantitative investigation was something a physicist could do. Many of my friends during the war had beat the engineers at their own game, and others had founded operations research. I

Rumination

tried to point out that the physicist's training is that of a scientific generalist and that there were many kinds of technical jobs they could do well, if they would just look beyond nuclear physics and quantum theory. In particular, they should be able to contribute in important ways to the problems of energy and food and pollution they orated so vehemently about.

By the time I became president, compromises had been reached on many of the most explosive issues. A poll of the membership had shown that a majority was willing to discuss the social implications of science at Society meetings, but was not in favor of the Society's becoming a lobbying organization. The majority felt that the Society's major task should still be the advancement and diffusion of the knowledge of physics. A new division of the Society, called the Forum, was established to organize sessions dealing with physics and public affairs. What seemed to be needed was to consolidate the new consensus and to remodel the committee structure appropriately.

This task took longer than my year's incumbency, but it was nearly completed by 1975. I found myself having agreed to chair the most recently formed committee, the Panel on Public Affairs, which was to oversee the Society's activities relating physics and physicists to public affairs. Physicists have useful things to say about many public problems, such as those of energy and of pollution, for example. The Society can initiate studies of aspects of these problems about which physicists can speak with authority, or it can provide experts from its membership to advise Congress or federal or state or urban administrations on these matters. Organizing such a committee so that it can respond quickly and effectively to opportunities as they arise has been an interesting and time-consuming task. It has reminded me at times of World War II days, when things had to be done quickly or not at all.

Now, in the second half of the seventies, I find myself between acts, as I was in the early fifties. I have eased out from under most of the tasks that kept me busy during the sixties. I took them on because they gave me a chance to explore, to learn, and to enjoy new ideas and experiences, and also because they gave me the chance to be in at the beginnings of new organizations that might contribute a modicum toward man's welfare. As with WSEG and Brookhaven earlier, I feel I have made my contribution to these later assignments. I can go back occasionally to visit them, nod appreciatively, and pat people on the back. But just now I am looking around for the next enterprise that needs help in getting started.

I cannot fool myself into thinking that I still am capable of doing creative research dealing with reality, not words. Since the sixties my tasks have been mostly those of persuading other, younger enthusiasts to do the creating, of pointing out where the cream was to be skimmed, and of persuading some agency to support their work. These tasks had to be carried out by someone, if progress was to be made; to me, though, they could never equal the excitement and deep satisfaction of the creative task itself. I did try to keep my supervisory span limited to a small number of individuals and the specific research they were doing, rather than climbing so high on the administrative ladder that individuals and realities would get lost in a cloud of words about general policies. I like to feel grass

Rumination

between my toes. It is clear I can no longer run up over the next rocky hilltop, but I can at least stand in the grass below and cheer the next generation on.

In retrospect, it is clear that the spurt of involvement I experienced in the sixties was not planned, nor was it actually as direct a progression of actions as the selectively condensed tale in the last two chapters might indicate. I undertook much of the work with only two general convictions, reached in 1955: first, that operations research could be helpful in many ways if people could be persuaded to use it and if enough experts could be trained and, second, that all educated young men should know how to program and use computing machines. Whenever an opportunity presented itself for me to further either conviction, I tended to act, to pull strings, to seek help from my many friends. If I felt I could do a job that would be worthwhile and fun to do, I preferred to go ahead and pull strings, rather than wait for someone else to take the initiative.

Of course, I tried to initiate many more undertakings than are mentioned in these chapters; these initiatives were not mentioned because the responses to them were inappropriate or absent. Only when I look back from the present can I see a steady progress, in spite of the detours and blind alleys.

It is also clear to me that fewer successes would have occurred had I not chosen, way back in 1923, to become a physicist. That choice, both in my training and in my friends, has made many things possible. My training has made me look facts in the face, made me want to measure them and to work out their implications, no matter whether they applied to atoms or automobiles. Because the small band of physicists were able to demonstrate their versatility in World War II and later, it turned out that nearly every time I looked for help in starting another project, I found a friend, or at least a friend of a friend, in some high place.

Many opportunities can still keep me busy. I can continue to try to persuade more physicists to think constructively about how their discoveries can be rationally applied and can assist these physicists in getting the ear of policymakers in government and industry. And I can try, by writing and talking, to persuade more young people to choose physics as a career.

During the fifties and sixties many students were attracted to physics by the allure of power and prestige, rather than by the promise of arduous labor that might satisfy their curiosity about the world. In the next twenty years, we may not need as many self-styled physicists as were turned out in the past two decades, but we will certainly need more than we turned out in the thirties. Only a small fraction of each generation is so endowed and motivated as to be able to enjoy the physicist's life. During the painful decades ahead we will need quite a few scientific generalists, those who can delve into, understand, and integrate the gamut of sciences, from neutrinos to DNA, who enjoy finding the linkages between new discoveries in all fields, and who are also willing to assist in working out the human implications of their discoveries. These generalists should be being recruited today, ten years before their training is to be finished. I suspect they will be attracted by the lure of intellectual adventure and the revelation of the coupling between mathematics and the real world, although these enticements are seldom used in our high

Rumination

schools or undergraduate colleges. These lures should be shown to the next generation more often and in more detail.

Recently I became, for a term, the chairman of the Governing Board of the American Institute of Physics, the organization that carries out the common activities of the related physics societies (optical, astronomical, acoustical, educational, research, and so on), such as publishing their journals and handling their public relations. It is not a new organization, of course, but there may be some new things it can do.

Of course I can continue to explore, if only at second hand. I can catch up on new progress in science; plate tectonics and exotic stellar bodies are two fields marked by exciting developments. Perhaps there will be a chance for me to use my analytic skills in some corner of these sciences. The two-volume *Methods of Theoretical Physics* needs to be rewritten; perhaps I can persuade Feshbach to collaborate again.

There are lots of places I have not visited; even without the excuse of organizing a conference, I might persuade myself to see some of them. A split cartilage in one knee forces an abatement of mountain-climbing; I can still outpace my son and son-in-law going up, but a rough trail down, particularly with a pack, gives me housemaid's knee. Less strenuous hiking, however, and bicycling and swimming are still feasible and enjoyable.

And, of course, I continue my reading diet of four or five books a week. In every succeeding decade enough new things are learned in archeology and history and enough new points of view are represented in biography to make the going back over earlier eras and lives almost a new exploration. I can spend a week playing records of Haydn symphonies or Beethoven quartets, or even take two weeks for all nine symphonies of Mahler. As I grow older, the slower movements are winning out. I find I turn more often to Schoenberg's *Verklaerte Nacht* or the cello parts of Strauss's *Don Quixote* or the slow movements of Mahler's Seventh Symphony or Mozart's no. 36 or Haydn's no. 101, than to any of Brahms or Tchaikovsky or even Stravinsky. And I ration my playing of the Mozart G Minor Quintet and Beethoven's Quartets, no. 8 and no. 14, lest the slow movements lose their marvellous flavor by my overindulgence.

Or I could turn frivolous again. Lord Rutherford's attributed statement, "Science is either physics or stamp collecting," reminds me that I could return to work on my collection of U.S. stamps. And I have not put together a coffee table or turned out a wooden bowl for more than ten years. Or I could go back to a different sort of collecting, genealogy. I once wrote, typed, and had reproduced a small book, for our children and near relatives, on the history of as many of their ancestors as I could find information about in this genealogy-rich city of Boston. Since then, many new records have been sent to me by people who had seen the book; some time I should put all this material together.

And I can write, as I am doing now, about science, about what it is like to do research in science, about what science can and cannot do for all of us. In my lifetime, more scientific advance has been made than in the whole previous history of mankind—an example of exponential growth. Scientific advance means an increase in our knowledge of nature and signifies a corresponding increase in the number of alternatives available for our choice. Of course, we can choose foolishly; we can seize one item out of the mass of new knowledge and transform it into an overspecialized technology that distorts our whole society—as we have done with the automobile. But that is not the fault of the knowledge; it is the fault of our choosing one small part without looking at the larger picture. The harm comes when some or all of us apply knowledge in misguided or selfish ways.

It is easy to say that if we and our leaders understood more of what is now known about nature, particularly about the interrelation of its parts, we could solve the problems of overpopulation, starvation, war, and pollution that now threaten us. But I am afraid the situation has gone beyond that. Greater appreciation of science by executives will, of course, help, but it will not help much until we solve the political problem of achieving a stable balance between individual freedom and the necessity to live within our common means.

The exponential has caught up with us. We can no longer have the individual freedom I enjoyed in my youth. I should not be allowed to shoot the last eagle or to cut down the last sequoia even though it may be on "my property." I should not be free to conceive a child if, as a result, that child or another has to die of hunger. A corporation should not have the right to deplete irreplaceable mineral resources or to damage irreversibly our environment. A city should not be allowed to foster segregation and pollution by its tax and zoning laws. A country should not have the right to reduce another country to poverty and starvation just to increase temporarily its own citizens' standard of living—or merely their dream of omnipotence.

We could be more free about those problems when I was young; we had more elbow room, the exponential had not caught up with us. Unfortunately, the habits and the common customs that worked fairly well then still govern us now, when they are inappropriate. We still feel we have the right to do what we wish with our own land, although the action may harm our neighbor. The Indian peasant still thinks it advantageous to have as many children as his wife can bear, although India cannot feed the children it has. The directors of a company still find it more advantageous to get what profit they can now, even at the expense of the environment or their workers' health—in fact, they would be forced to resign if they did otherwise. A head of state still finds it profitable politically to enrich his country at the expense of another, poorer nation—in fact, he would not long be head of state if he did not at least try it.

We can no longer afford these kinds of freedoms. The world is becoming one huge city, and urban controls have always had to be more restrictive than rural ones. But who applies the controls? Is the only answer a world tyranny, in the original Greek sense? This is the political problem that must be solved, or it really will turn out that the human race has fatally infected the world.

The ironic fact is that we are just now starting to know about the world as a system. We might even now reverse the exponential if only that political problem were solved. Ecology, systems analysis, and economics are emerging pieces of a science of world interrelationships that promises to discover alternatives to the old solutions of war, famine, and plague—if it can be made advantageous for people to cooperate for the long pull, rather than to compete for short-term gains.

Any system needs control devices if it is not to destroy itself. The feedback controls of the social system require long-term planning, investigation of the side effects of each possible major development, determination of the possible consequences of each of the various alternatives handed us by science, before we choose which to apply, and then establishing incentives and prohibitions so that the development goes in the desired direction.

Planning requires measurement of the system, particularly of the interactions between its parts. The results of measurement must then be expressed in a set of equations or a computer program that will simulate its behavior well enough to enable us to predict the results of alternative policies sufficiently well to be able to choose among them. This planning is not a repudiation of human values; it is an attempt to save them. The rich and valuable truths of humanism are concerned with individuals. The methods and the truths needed to plan, to present clearly the alternatives, must be concerned with masses of people and resources, must be the methods of ecology, of operations research, of science in general. No conflict exists between the two; they complement each other, for the choice must in the end be made in accord with human ideals and desires.

Long-range planning is beginning in many countries. It began first in Europe because of Europe's longer background of centralized direction. Only recently has the word "planning" lost its ominous connotations in this country, but now even here the pressure of events is forcing a change of attitude. We must see more systems planning at the federal level, as well as at the state and local levels.

The weakest point at present is in international controls. Within each country, social destruction is beginning to be avoided by a network of restraints and incentives, guided as yet only clumsily by long-range planning. Just as individual autonomy must be restricted, so national autonomy must be reduced, to enable international planning to be more than an academic exercise. How much reduction is absolutely necessary and how the needed controls are to be administered and enforced are the most urgent problems we face today.

Many popularizers of science write about the marvellous things that could result from the application of this or that new scientific discovery, but they seldom speak of the side effects, the costs of such application. And they neglect to stress the basic problem, that without plans and controls the tempting new application could turn into another catastrophe.

In the end someone—either one person or a few or all of us—will have to choose among the many alternatives. The professional planners should not make the decision. Their task is first to work out the consequences of each possible choice and to present them, clearly and explicitly, so people can decide, then to devise rules and incentives to implement the decision, and then to find ways to measure how close the result comes to what had been desired. The task will never be simple or clear-cut, but it must be accomplished if we are to avoid the old solutions: war, starvation, and pestilence. These global political problems of choosing and then enforcing the choice are beyond the scope of science, but science is needed for their solution.

The answer, of course, is not to curtail science, for science presents new possibilities, and possibilities are never dangerous by themselves. Nor is the answer to channel research more stringently; as my life demonstrates, one never knows ahead of time what useful knowledge will result from any research project, no matter how impractical or dangerous it may seem at first. Of course, money for research is not limitless, so a rough set of priorities will have to be assigned, to be changed from year to year. But I am convinced that coupling these priorities too closely to short-term needs is a quick way to dry up those long-chance breakthroughs that can appreciably add to our alternatives for action.

But philosophizing brings few converts. My task in this narration has been to tell my story as frankly as possible, in the hope that the reader can sort out the pattern and the message. Of course, to paraphrase Solon's reputed comment to Croesus, no autobiography can have an artistically satisfying conclusion; the end is not yet. However it seems to me that a few patterns emerge and a few lessons can be learned from the life of one of the first wave of home-grown American scientists. For those who like exploration, immersion in scientific research is not unsocial, is not dehumanizing; in fact, it is a lot of fun. And, in the end, if one is willing to grasp the opportunities, it can enable one to contribute something to human welfare.

Name Index

Abramowitz, Milton, 284
Acheson, Dean G., 258
Adams, Edwin P., 58, 70
Allis, William P., 99, 100, 107, 108, 114, 119, 121, 136, 147, 213
Archambault, Bennett, 192
Astin, Allen V., 284

Babcock, Gen. Allen, 327
Bacher, Robert F., 104, 128
Bainbridge, Kenneth T., 56, 62, 72, 99
Baker, Capt. Wilder, 172, 174, 180, 182, 189
Barberi, T., 345
Barnes, Gen. Earl, 245, 248
Barrett, Richard L., 13, 22, 38
Barrow, Wilmer L., 136
Beer, Stafford, 345
Bell, Maurice E., 180, 207
Beranek, Leo L. 144, 160, 219
Bethe, Hans A., 149-150, 232
Bhatt, Nautam, 145
Bitter, Francis, 99
Blackett, P. M. S., 114, 172, 189, 192
Blewett, John P., and M. Hildred, 227, 236
Bloch, Felix, 114
Bohr, Niels, 71, 78
Boiteux, Marcel, 323
Bolt, Katherine, 144
Bolt, Richard H., 144, 156, 161, 219-220, 264
Born, Max, 78
Borst, Lyle B., 224, 236
Bowles, Edward L., 120, 187, 213, 240

Bradley, Gen. Omar, 248, 258
Bragg, William L., 111
Brode, Robert B., 87, 128
Bronck, Detlev W., 339
Brown, Arthur A., 206, 290
Brown, Gordon S., 352
Bush, Vannevar, 121, 147, 154, 170

Casals, Pablo, 103
Chandrasekhar, S., 152
Cochrane, C. A., 337, 345, 347
Cockroft, John, 114
Coffinberry, Arthur, 23, 27
Collbohm, Frank R., 240
Compton, Karl T., 55, 63, 65-66, 70-71, 76, 92, 99-100, 118, 125, 127, 134, 154, 161, 171, 206, 214, 249, 282
Compton, Margaret, 99
Conant, James B., 154, 170
Condon, Edward U., 82, 86, 98-99, 129, 280
Conrad, Capt. Robert H., 216
Corbato, Fernando, 308, 351
Craig, Cmdr. E. C., 156, 161, 162

Dammerstadt, Gunnar, 345
Darrow, Karl K., 65, 95
Davis, Bergen, 89
Davisson, Clinton J., 92, 96-97
Dent, Col. Fred C., 159
Deutsch, Martin, 215
Dobbie, James M., 332
Douglas, Donald, 240
Draper, C. Stark, 135, 144

Eberstadt, Ferdinand, 238, 244

Name Index

Ehrentest, Paul, 78-79
Einstein, Albert, 43, 71, 87, 232
Eisenhart, L. P., 58, 71, 99

Falch, Francisco, 325
Fay, Richard D., 120, 123, 137, 143, 161
Feld, Bernard T., 215
Fermi, Enrico, 102-104
Fermi, Laura, 102
Feshbach, Herman, 133, 215, 218, 221, 243, 269, 272-273, 336, 356, 363
Feynman, Richard P., 126, 135
Fisk, James B., 135, 213
Forrestal, James V., 211, 238, 244, 249
Forrester, Jay W., 276, 300, 313
Fowler, Hugh, 342
Fox, Milton, 14
Frank, N. H., 119
Frasché, Dean F., 339
Fraser, Lindlay, 57
Frisch, David H., 215, 309

Gaither, H. Rowan, 241
Galliher, Herbert P., 295-296, 315, 321, 324, 328
Garber, Newton, 293
Gavin, Gen. James M., 245, 248
Germer, Lester H., 93, 97
Goertler, Henry, 326
Goodeve, Sir Charles, 299, 326, 329, 336
Goudsmit, Samuel, 78-79, 102
Groves, Gen. Leslie, 221-223
Gurney, Ronald W., 58, 228

Hanson, William L., 128, 133
Harnwell, Gaylord P., 92
Harrison, George R., 119
Haworth, Leland J., 243
Heisenberg, Werner, 78, 81, 114
Heitler, W., 83
Helly, Walter, 315
Hitler, Adolf, 106, 112, 150, 153, 157
Hopkins, Arthur M., 94-95

Hopkins, James, 165
Hopkins, Martin L., 94
Hopkins, William R., 94
Hovarth, William J., 248, 252, 259
Howard, Ronald A., 316, 321, 324, 332
Hsieh, Hilda, 273
Hubbert, M. King, 339, 341
Hull, Gen. J. Edward, 244-245, 248, 255
Hunt, F. V., 143, 160
Hurd, Cuthbert, 303

Idei, Shigeru, 335
Ingard, K. Uno, 356

Jacob, Col. Itzak, 348
Jennings, John B., 316-317
Johnson, Louis A., 249, 258
Johnson, Ralph P., 135

Kapitsa, Peter, 114
Kelly, Mervin J., 97
Kemble, Edwin C., 87, 127
Kennedy, Pres. J. F., 339, 342
Killian, James R., 278
Killian, Thomas J. 62, 99, 294
Kimball, George E., 128, 133, 183, 210, 238, 248, 290, 321, 324
King, Adm. E. J., 171-172, 182-183, 184, 190, 211
Kip, Arthur F., 180-181, 207, 209
Kistiakowski, George B., 98
Klippenberg, Eric, 325-326, 348
Koopman, Bernard O., 206, 319-321, 324, 326, 330-331
Koster, George F., 350
Koyonagi, Kenichi, 329-330, 333-334, 348
Kramers, H. A., 78-79, 129
Kuper, J. B. H., 225
Kuper, Mariette, 98, 225

Langmuir, Irving, 64, 74, 89

Name Index

Larkin, Gen. Earl, 320-321, 323, 327
Laukhuff, Richard, 26, 28
Lawrence, Ernest O., 224, 231
Leach, W. Barton, 192
Lee, Adm. Willis, 209
Lemaître, Abbé Georges, 122, 339
LeMay, Gen. Curtis, 255, 258-259
Levinson, Horace C., 290
Lied, Finn, 325, 328
Little, John D. C., 296, 355
Livingood, John J., 61, 73, 99
Livingston, M. Stanley, 207, 224, 237
London, F., 83
Lorentz, H. A., 87
Low, Adm. F. S., 184-186
Lowan, Arnold, 139, 274
Lowell, Capt. Robert, 162

Mack, Julian E., 99
Magie, W. F., 60-61
Martin, John R., 52
Massey, Sir Harrie, 115
Matsuda, Takehiko, 335
McBane, Louis, 30-31, 34
McCarthy, John, 305, 307-308, 312
McCarthy, Sen. Joseph, 280, 283, 343-344
McCarthy, Philip J., 180
McCord, William B., 12-13
McCord, William C., 60, 68, 70
McNamara, Robert S., 260
Menzel, Donald H., 147, 152
Michelson, Albert, 43, 45
Miller, Dayton C., 34, 37, 40, 43, 45-46, 48, 50, 54-56, 82-83, 120, 123, 157
Millikan, R. A., 134
Mills, Adm. Earle, 239
Minsky, Marvin L., 305, 307-308, 312
Morley, F. W., 43, 45
Morris, Joseph C., 62, 72-73, 80-81, 92, 99, 337

Morris, Lloyd W., 52-53
Morse, Allen C., 9, 10, 13, 18, 21, 30, 38, 54-55, 69, 116, 138
Morse, Annabella, 153, 342
Morse, Annabelle H., 91-92, 94-95, 97, 100, 135, 222, 229, 243
Morse, Conrad P., 222
Morse, Edith F., 12-13, 21, 38, 138
Morse, John Flavel, 8
Morse, John Franklin, 10, 23
Morse, John Franklin, Jr., 8
Morse, Marston, 290
Morse, Richard W., 13
Mothes, Jean, 323, 345
Mott, Neville, 114

Nassau, Jason J., 25-26, 28, 36, 42, 49, 51, 55, 69
Nierenberg, William A., 331
Nixon, Sen. Richard M., 281
Notestein, Frank W., 339
Nottingham, Wayne B., 119
Nygaard, Kirsten, 325

Olshen, A. C., 183-184, 210
Oort, J. H., 51
Oppenheimer, J. R., 87-88, 95, 250, 283
Owens, Dean, 52

Palmer, A. Mitchell, 28
Parmenter, Lt. Richard, 162
Parsons, Adm. W. S., 211, 245, 247
Partridge, Bellamy, 95
Pauling, Linus, 111, 232
Pauling, Helen, 111
Pegram, George B., 221
Pellam, John R., 180, 205
Peter, J. Georges, 225-226
Piel, Gerard, 234
Planck, Max, 71, 78
Pompilj, Giuseppe, 327, 332, 334
Porter, Frank, 22, 25

Name Index

Rabi, I. I., 215, 328
Rabinowitz, Eugene, 234
Ramsey, Norman F., 221, 224, 319
Randall, H. M., 78
Revans, R. A., 346
Revelle, Roger, 339
Rickover, Adm. Hyman, 239
Rinehart, Robert F., 180, 205, 207
Roberts, R. L., 19, 21, 30, 38
Robertson, H. P., 82-83, 98, 193-194, 259
Roderick, Hilliard, 337
Rosentiehl, Pierre, 322
Rosseland, S., 147-148
Rosser, J. Barkley, 338, 342
Rossi, Bruno, 215
Russell, Henry Norris, 71-72, 146, 151-152
Rutherford, Sir Ernest, 114, 116, 364

Sabine, Wallace, 143-144
Salzmann, Charles, 299, 323, 326
Sasieni, Maurice W., 329
Schaffner, Joseph, 232
Schiff, Leonard I., 135
Schroedinger, Erwin, 78, 81
Seitz, Frederick, 319, 323-324, 326, 328, 331
Shankland, Roberts, 50
Shapley, Harlow, 147, 152
Shockley, William, 135, 182, 187, 189, 194, 213, 260
Sidis, William, 140-141
Slater, John C., 98, 100, 119, 213
Smyth, Harry D., 58, 76, 210
Sommerfeld, Arnold, 78, 100, 105, 107-108, 110
Spannyonakis, Gen. R., 336
Stegun, Irene A., 284
Steinhardt, Jacinto, 185-186, 205
Stevens, S. S., 160

Stoller, David S., 332
Straight, Michael W., 232
Stratton, Julius A., 101, 119, 213, 278
Stueckelberg, Ernst C. G., 74, 80-81, 88, 100, 107, 112, 114, 117
Swart, Lawrence A., 225
Szilard, Leo, 232, 234-235, 280

Tada, Col. Kazuo, 333, 336
Tate, John T., 86-87, 171, 173
Tate, Vernon D., 288
Taylor, Hugh S., 98
Teager, Herbert E., 308
Teller, Edward, 84, 114
ter Hart, W., 347
Thiesmyer, Lincoln R., 225
Thomas, Charles D., 77-78, 80
Truman, Pres. Harry S., 258
Turner, Louis A., 58, 84

Uehling, Edwin A., 203
Uhlenbeck, George E., 78-79, 102
Urey, Harold C., 232
Uyterhoeven, W., 74

Vallarta, Manuel S., 101, 119, 121-122, 339
van de Graaff, Robert, 73, 101, 119
von Neumann, John, 87, 98, 103, 129, 274, 301
von Neumann, Mariette. *See* Kuper, Marriette

Wadsworth, George P., 297, 321, 324
Warren, Bertram E., 101, 119
Waterman, Alan T., 283
Watson, Thomas J., Jr., 304
Weiss, Paul, 339
Weisskopf, Victor F., 215, 232
Welsh, George I., 248, 259
West, Andrew F., 56, 77
White, J. Edward, 162

Name Index

Wiener, Norbert, 140
Wigner, Eugene P., 98
Williams, Gordon, 32, 54
Williams, Hope, 95
Williams, John D., 316
Wood, R. W., 89

Yaffe, Harold J., 318
Yilmaz, Huseyin, 266
Yngve, Victor, 312
Young, Lloyd A., 128

Zacharias, Jerrold R., 215, 221–222
Zuckerman, Sir Solly, 328
Zurcher, Arnold, 56–57